The Physics
of
Vibrations
and
Waves

David M. Scott

The Ohio State University

Merrill Publishing Company
A Bell & Howell Company
Columbus Toronto London Sydney

to Pat

Cover Photo: Fundamental Photographs

Published by Merrill Publishing Company
A Bell & Howell Company
Columbus, Ohio 43216

This book was set in Century Schoolbook.

Administrative Editor: Robert Lakemacher
Production Coordinator: Jeffrey Putnam
Cover Designer: Cathy Watterson

Library of Congress Catalog Card Number: 85-61996
International Standard Book Number: 0-675-20396-1
Printed in the United States of America
1 2 3 4 5 6 7 8 9—91 90 89 88 87 86

Preface

Why can we hear, but not see, a person around the corner of a building? Why do we often see red sunsets or sunrises while the sun is rarely red at noon? Is there a physical reason for the surfer's truism that every seventh wave is a big one? Why do distant mountains have a bluish tinge? These and hundreds of other questions can be answered using the single concept of waves. Because they help us understand so much of the world around us, waves are worth studying.

This book is about waves: how they form, how they move, and how we detect and describe them. Vibrations are included because a wave is the transportation of vibrations from one location to another. The many physical phenomena explained in this book illustrate general wave behavior. The examples involve sound, light, or water waves because most people can relate to them so easily.

This book is designed as a text for a one-term course for students who need an introduction to physical science and who have very little preparation. Rather than using higher mathematics, which some students may not be ready for, I use diagrams and simple graphs to describe the physics and extend the ideas presented.

I have two caveats for using this book. First, vibrations and waves need movement to be fully appreciated. Student understanding of the book will be greatly enhanced by experiencing the events described in the book through demonstrations, films, etc. It is better, of course, if they have the opportunity to perform simple experiments—both qualitative and quantitative. The book is not designed to stand alone. Second, students must recognize the subject matter covered occupies only a small fraction of that offered in a general physics course. Because the breadth of the subject is limited, we can delve more deeply into it.

The first five chapters deal with vibrations and the production and detection of sound. Chapters 2 and 3 introduce the vocabulary for discussing simple and complicated vibrating systems. The aspect of sound we call music is introduced in chapter 4, the human ear in chapter 5.

Chapter 6 presents the subject of waves. Various aspects of waves are covered in the remaining chapters: water waves, musical instruments, light waves and the eye, the laser, holography, and quantum mechanics.

This book is based on a course developed by the physics department at the Ohio State University to provide an exposure to physics for students who are not studying for degrees in the physical or life sciences. Business, education, social science, humanities, and art majors continue to enjoy and benefit from the course. Many of these students had their last experience with the physical sciences in the ninth grade. These students are often fearful of mathematics and intimidated by physics.

During its beginning stages, the course development was guided by a committee of interested faculty and graduate students. Many other people have added their ideas to the continuing development of the course, and I have benefited from all these contributions. To these people I owe a debt of gratitude for their implicit or explicit help in developing this book. I can't mention them all by name, but I must single out the course-development coordinating committee: Gene Ferretti, Bernie Mulligan, Noel Stanton, Kai Tanaka, and Josanne Trusty, as well as Len Jossem, who was department chairman at that time and provided guidance for the project. The people at Merrill Publishing Company have been very helpful during the publishing process; I certainly must mention Bob Lakemacher, Mark Garrett, Cathy Watterson, Jeff Putnam, and especially Susan King. Finally, Mrs. Babette Mullet who typed numerous drafts of the book deserves a gold medal!

David M. Scott

Contents

1. **Introduction to the Study of Waves and Vibrations, 1**

   An Overview of Waves and Vibrations, 2
   The Observation of Sounds on the Oscilloscope, 7
   Lissajous Figures, 9

2. **Simple Vibrations, 17**

   Description of Vibratory Motion, 18
   Experimental Uncertainty (A Digression), 20
   Sine Curve Readings, 21
   Phase of Vibratory Motion, 22
   Damping, 24
   Resonance, 25
   Coupling, 28
   Other Examples of Resonance, 29
   Measurement of Rapid Vibrations, 31
   Other Vibrations, 34
   Superposition and Beats, 34

3. **Complicated Vibrating Systems, 43**

   Normal Modes, 44
   Superposition of Normal Modes, 50
   Vibration Recipe, 54
   Starting a String Vibrating, 55
   Vibrating Surfaces, 59
   The Production of Sound by a Loudspeaker, 60

4. **The Piano as a Source of Sound, 69**

   The Piano as a Vibrating System, 71
   Musical Scales, 77

5. The Ear as a Detector of Sound, 89
    Capability of the Ear, 90
    Structure of the Ear, 97

6. Introduction to Waves, 105
    Traveling Waves on a String, 107
    Transverse and Longitudinal Waves, 111
    Sound as a Longitudinal Wave, 112
    Water Waves, 114

7. Ocean Waves, 127
    Sources of Ocean Waves, 128
    Surf Beat, 134
    The Breaking of Waves, 135
    Mass Transport, 137
    Funneling Effects, 140

8. Standing Waves, 147
    Wave Reflection at the End of a String, 148
    The Production of Standing Waves, 152
    Normal Modes as Standing Waves, 157
    Organ Pipes, 163

9. Examples of Standing Waves, 173
    Seiches, 174
    Stringed Instruments, 178
    Wind Instruments, 183
    Woodwinds, 183
    Brass Instruments, 187

10. Light as a Wave, 193
    Description of Light Waves, 194
    Color, 198
    Sources of Light, 202

11. Wave Movements, 209
    Wavefronts and Rays, 212
    Refraction, 214
    Reflection, 217
    Total Internal Reflection, 218
    Focusing of Waves, 222
    Simple Lenses, 226
    Images, 229
    Dispersion, 233
    The Doppler Effect, 236

12. The Eye as a Detector of Light Waves, 247

Similarities Between the Eye and a Camera, 248
Describing Light, 249
Capability of the Eye, 253
The Structure of the Eye, 255
Eye Defects, 258

13. Other Wave Behavior, 265

Diffraction, 266
Interference, 271
Scattering, 281
Polarization, 284

14. Atmospheric Phenomena, 299

The Rainbow, 300
Scattering Effects, 304
Halos and Coronas, 306
Sun Pillars, Sun Dogs, and Sub Suns, 309
Other Atmospheric Phenomena, 311

15. Lasers and Holography, 317

Lasers and Light Sources, 318
Uses of Laser Light, 322
Holography, 324

16. Epilogue 333

Glossary, 337

Answers to End-of-Chapter Questions, 340

Index, 341

About the Author, 344

1

Introduction to the Study of Vibrations and Waves

Important Concepts

- Waves carry energy from one location to another without transporting any matter.

- The energy transported by waves occurs as vibrations.

- Most waves require a medium through which to move; light waves (and other electromagnetic waves) are the exception.

- Two types of curves can be generated (and observed with an oscilloscope) by vibrations. The vibration pattern (or waveform) shows how the position of the vibrating object changes over time, and two vibrations can produce a Lissajous figure.

An Overview of Waves and Vibrations

The subject of waves at first might seem to be a very narrow topic because we all associate waves with water moving across the surfaces of lakes and oceans. If this movement of water were only part of the physical world that could be explained using the concept of waves, then I would not be writing this book. But the view of waves is much broader. In fact, physicists have applied their understanding of waves to explain situations ranging from the behavior of subatomic particles to the size of the universe. In this book, I take a middle ground approach when considering the breadth of the waves topic by looking at three different areas of the world around us where wave concepts apply: water, sound, and light.

One of the beauties of science and, indeed, knowledge in general is that insight can be transferred from an area where understanding is well developed to another area where confusion still exists. Of course, this does not happen just in physics. Psychologists study animal behavior to better understand human behavior; sociologists study isolated societies to better understand modern culture; and anatomy students dissect cats to better understand how the bodies of mammals function. In this book, we will take a similar approach. That is, we will develop some aspect of the waves subject by looking at a particular example. Then, we will transfer the understanding gained through that particular example to other situations, and we will stress the similarities and differences that exist among the cases being discussed. For example, when we first discuss the movement of waves from one place to another, we will do so by considering waves on strings and bodies of water. Then, we will use the ideas developed there to help us understand sound waves and light waves.

Before we look at things that happen in the world around us that can be explained by waves, we must first understand how waves work. If we watch a wave moving across a lake, and if we concentrate on watching one point on the surface of the water rather than allowing our eye to follow the wave as it moves across the surface, then we will notice that as the wave passes that point, the surface moves up and down. If the wave is smooth and regular, the surface moves up and down in a uniform type of motion (called *vibration*).

Another way of describing this relationship between vibration and wave is to say that a wave is the transporting of vibrations from one place to another. It is difficult to understand and describe waves without knowing about vibrations, so this book

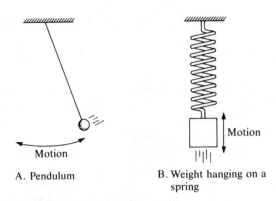

Figure 1.1. Simple oscillators

A. Pendulum

B. Weight hanging on a spring

will focus on vibrations first. We will begin our study by looking at *simple vibrating systems,* or to use another term, *simple oscillators.* Because it is helpful to be able to see the vibrations of an oscillator, we will first study mechanical vibrating systems such as a pendulum or a weight hanging on a spring (Figure 1.1). After studying simple oscillators, we will move on to consider more complicated vibrating objects such as strings and plates. In addition to considering examples of oscillators (which we can visually observe), we will look at the production of sound by means of vibration. Vibrating objects producing sound can be simple oscillators or more complex systems. We will look at various musical instruments as examples of complicated vibrating systems that produce sound.

At this point in our study, we will be prepared to consider waves. As we just mentioned, a wave is the concept we use to describe the transferring of vibrations from one place to another. A wave originates at one location from a wave source (which is an oscillator), travels through space, and gets detected at another location by a detector (which is another oscillator). This process is illustrated for one case in Figure 1.2 which shows the relative locations of the source and detector for a sound wave.

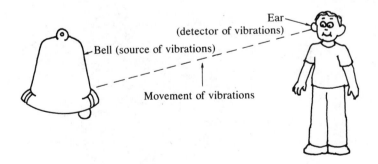

Figure 1.2. The movement of vibrations (by means of a sound wave)

With sound, we cannot see either the vibrations or the waves. The same problem exists for light. Thus, we first will look at waves we can see such as ones that travel on strings or across bodies of water.

Vibrations contain energy, and since waves transfer vibrations from one place to another, it follows that waves carry energy through space. There are other ways to transport energy. For example, a stone that is thrown will carry energy with it. Waves are unique, however, because they move energy from one place to another without moving any matter. An interesting comparison of two energy transportation methods can be found in the transfer of information (which requires the transfer of energy). If you want to tell a friend in a distant city some news, you can either call that person on the telephone or write a letter. When you convey the information by means of a call, you use waves that travel through the wires to send the energy (Fig. 1.3). The letter, however, has the information attached to matter, and both the energy and the matter are transported together from one place to another.

As most waves carry vibrations across space, they move through a *medium,* which is some type of material or matter. When a wave moves along a string, for example, the medium through which the wave travels is the string itself. The medium for water waves is the water, and light waves travel through air, water, glass, and other transparent materials. Sound waves also can move through various media such as air, water, or earth. An interesting example of the latter occurs in old cowboy movies where someone puts his ear to the ground to hear the sound of approaching hoofbeats.

A standard demonstration in an introductory physics course involves placing a ringing bell inside a container, evacuating the

Figure 1.3. The movement of vibrations during a telephone conversation

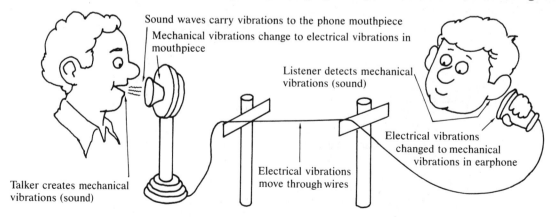

Sound waves carry vibrations to the phone mouthpiece

Mechanical vibrations change to electrical vibrations in mouthpiece

Listener detects mechanical vibrations (sound)

Electrical vibrations changed to mechanical vibrations in earphone

Electrical vibrations move through wires

Talker creates mechanical vibrations (sound)

container, and observing that the sound disappears. From this, we conclude that the sound travels through the air, but not through a vacuum. When we perform the same experiment with a light bulb inside the container, we find a very different result: As the air is removed, the light we see coming from the bulb does not change! Thus, we discover that light can travel through a vacuum. We should not be surprised by this result because light emanating from the sun and stars travels through the emptiness of outer space to reach the earth. This light bulb experiment emphasizes a unique characteristic of electromagnetic waves: They do not need a medium to travel. (Light waves are just one form of electromagnetic waves. Other types of electromagnetic waves include ultraviolet light, infrared radiation, and radio waves.) All other waves need some type of medium through which to move.

When it reaches its destination, a wave is detected. We often use our eyes or ears to detect waves directly in the world around us. At other times, however, we use instruments to detect waves, either because the waves cannot be detected by human senses or because our understanding of the waves can be increased.

A bell inside an evacuated enclosure (a bell jar) sitting on top of a vacuum pump

The Oscilloscope

An important instrument used to help us visualize and study the vibrations and waves of sound is the oscilloscope. If we hook up a microphone to an oscilloscope, we can use the oscilloscope to "look at a sound wave." Let me describe how this process works. First, the microphone detects variations of pressure in the air and produces electrical signals in response to these pressure variations. For sound, these variations are vibratory and are carried to the microphone from the sound source by means of a wave. The electrical signals from the microphone are transmitted to the oscilloscope which then shows the pattern of the vibrations on a screen. When we talk to someone on the telephone, the vibrations are changed in a similar way. Talking creates mechanical vibrations in the air. Within the mouthpiece of the telephone, a microphone changes them into electrical vibrations, which travel through the telephone line and are changed back into sound vibrations by the earphone in the instrument of the listener.

The vibration pattern picture which appears on the oscilloscope screen is actually drawn by a dot moving rapidly across the screen. The dot moves horizontally across the screen at a con-

An oscilloscope with a sine
curve displayed on its screen

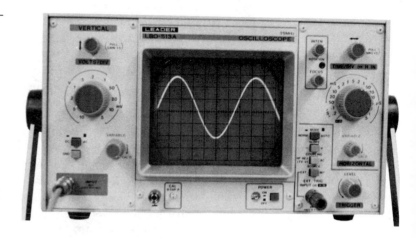

stant speed and moves vertically in response to the vibrations
that it is picturing. To understand this, first suppose there is no
vibration. Then, the dot moves straight across the screen, and as
soon as it completes one sweep across the screen, it starts another. The oscilloscope has a control to change the horizontal
speed of the moving dot. When the dot moves slowly, we can
easily follow its motion, and no line is visible on the screen. But
at higher speeds, the dot moves so rapidly that it appears to draw
a line. This line occurs for two reasons. First, an image of the
moving dot remains at each spot it touches for a short period of
time after the dot has passed. Second, as we will see in Chapter
12, the eye has persistence of vision. For a rapidly moving dot,
these two effects combine so the oscilloscope displays a smooth
line. When there is no vibration, the line is straight and horizontal.

When there is a vibration, the dot not only moves horizontally as described, but it also moves vertically. The vertical
motion of the dot on the screen corresponds to the movement of
the vibration. Figure 1.4 shows how this works for a simple oscillator that is constructed from a mass (that is, a weight) hanging
on a spring. Simple oscillators will be discussed in greater detail
in Chapter 2; here, we only need to note that during its vibration, the mass moves repetitively up and down. With suitable
electronics, the position of this mass can be detected and displayed on an oscilloscope screen, where the vertical movement of
the mass creates a corresponding vertical movement of the dot.
Unlike the mass, however, the dot moves horizontally across the
screen at the same time it is moving vertically in correspondence
with the vertical motion of the mass. The resulting picture on

A spring-mass system
consisting of a weight (mass)
hung on the end of a spring

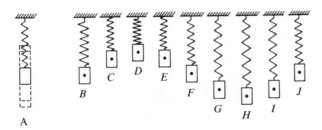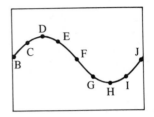

the screen is a smooth curve, which we call a *vibration pattern*. Physicists also call this curve a *waveform* because the vibrations are being carried by a wave.

Figure 1.4. A vibrating system made from a mass suspended on a spring. The resulting trace on an oscilloscope screen is also displayed.

This correspondence between the motion of the vibrating mass and the picture drawn on the oscilloscope screen is shown in Figure 1.4. To understand this figure, we must realize that the mass on the end of the spring moves only up and down. Since it doesn't move horizontally, pictures of it at different times would lie on top of one another as shown in A. Such a picture is obviously difficult to understand, so we put the pictures that occur at different times side-by-side as in B–J. Each of these pictures shows the position of the mass at a different time, but the horizontal position of the mass is not changing. The time interval from one picture to the next is the same in each case.

The correspondence between the moving mass and the vibration pattern on the screen is seen quite easily. In picture B, the mass is located at the midpoint of its motion and is moving upward. The corresponding point on the oscilloscope screen also is marked *B*. A short time later, the mass has moved to the location shown in picture C, and the dot on the oscilloscope screen has moved to the point marked *C*. During the time between pictures B and C, the mass moves continously upward, and the dot traces out the line between points *B* and *C*. During the subsequent time intervals when the mass changes its location, as shown by pictures D through J, the dot on the screen traces out the curve shown. As will be discussed in Chapter 2, this spring-mass system is called a *simple oscillator,* and the pattern on the oscilloscope screen is a particularly simple curve called a *sine curve.* For a more complicated vibrating system, the vibration pattern recorded on the screen of the oscilloscope can become correspondingly more complex.

The Observation of Sounds on the Oscilloscope

When the oscilloscope is connected so we can observe sound vibrations, we can make some observations about how the pat-

Figure 1.5. Some vibration patterns on an oscilloscope screen showing effects of changing pitch and loudness

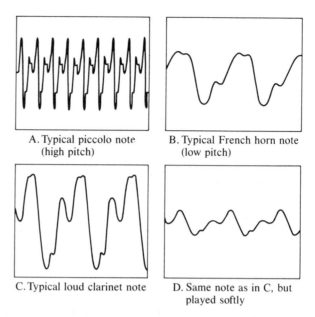

A. Typical piccolo note (high pitch)

B. Typical French horn note (low pitch)

C. Typical loud clarinet note

D. Same note as in C, but played softly

terns displayed on the screen vary with changes in the sound. (I am assuming the oscilloscope settings are not being changed during the course of these observations.) The first observation we can make is that in general the wiggles on the screen are closer together for high-pitched sounds than for low-pitched ones. For example, the sound of a foghorn causes wiggles that are farther apart than the sounds of a soprano singing an aria. And, a piccolo's sound produces wiggles much closer together than that of a French horn. The second observation we can make is connected with the loudness of a sound. Generally, the louder a sound, the taller the wiggles on the screen. We will see in Chapter 5 that this observation is not always true, but at first glance, it certainly seems true. Several vibration patterns for different sounds are shown in Figure 1.5.

When we look at sounds produced by some objects—like a tuning fork—we find the resulting oscilloscope wiggles are very uniform in size and shape; a sine curve vibration pattern forms as shown in Figure 1.6. These objects are simple oscillators and are discussed in Chapter 2. Those sounds producing more complicated vibration patterns are discussed in subsequent chapters. Most sounds produced by musical instruments fall into the latter category. Musicians classify sounds with a sine curve vibration pattern as dull and uninteresting. Rich, full-bodied tones always will have more complicated vibration patterns.

Two special variations of musical notes that deserve particular mention are *vibratto* and *tremulo*. Vibratto is a rapid

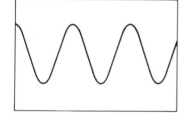

Figure 1.6. The vibration pattern of a tuning fork: a sine curve

variation in the pitch of a note being sung or played on an instrument. Thus, when vibratto occurs, the wiggles on the oscilloscope screen change their separation rapidly. As the pitch increases and then decreases, the wiggles move closer together and then farther apart. Tremulo, on the other hand, is a rapid variation in the loudness of a note. On the oscilloscope screen, this variation will appear as a rapid change in the height of the curve.

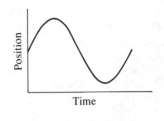

Figure 1.7. A graph of position versus time for the oscillating mass of Figure 1.4

Lissajous Figures

As we have seen, a vibration pattern on an oscilloscope screen shows how a vibrating object changes its position with time. In fact, the pattern can be interpreted as a graph of position versus time, with position plotted vertically and time horizontally. Figure 1.7 shows the position versus time curve for the oscillator in Figure 1.4. This is a sine curve because, as we noted, the spring-mass system is a simple oscillator.

It is possible, however, to produce a different type of picture on the screen of an oscilloscope. Such a picture, called a *Lissajous figure* after Jules Antoine Lissajous, is produced when the oscilloscope is connected to two different vibrating systems. As before, one of the vibrating systems causes the dot to move vertically on the oscilloscope screen. Now, however, the other vibrating system causes the dot to move horizontally back and forth rather than across the screen at a constant rate. When the dot vibrates simultaneously in both directions, the resulting pattern can be a circle, an ellipse, a diagonal line, or a much more complicated picture.

As noted before, the first oscilloscope display we discussed (called a *vibration pattern*) shows how the position of a vibrating object varies with time. A Lissajous figure does not give the same information. Its shape depends upon how the two vibrations compare with each other. Figures 1.8, 1.9, and 1.10 show how the two vibrations produce this pattern. In viewing these figures, it is important to recognize that (just as in Figure 1.4) the different pictures of the oscillators occur at different times. (The confusing overlayed picture has been omitted.) Thus, in each figure, the two sequences of pictures show the movements of two masses hooked to springs. In each sequence, the picture labeled by the same letter occurs at the same time; the same letter also indicates the resulting location of the dot on the screen. For example, in Figure 1.8, at the time indicated by *B*, the vertically oscillating mass is partway down from the top, and the horizontally oscillating mass is partway to the right end of

Figure 1.8. The production of a Lissajous figure with two vibrating systems. Both systems have the same frequency, but they are oscillating 90° out-of-phase with each other.

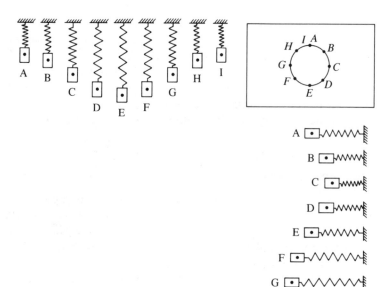

Figure 1.9. The production of a Lissajous figure with two vibrating systems. Both systems have the same frequency, and they are moving in phase with each other. (Note the horizontally oscillating mass starts differently in Figures 1.8 and 1.9.)

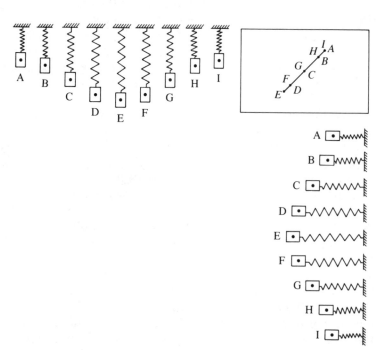

its motion. Then at C, the mass moving vertically has reached the center of its motion (and is still moving down) while the mass moving horizontally is at the extreme right end of its motion. Thus, the location of the dot (which is controlled by the positions of the two masses) is at the right side of the circle. Further movement of the dot around the circle is caused by the repetitive motion of the masses.

In Figures 1.8 and 1.9, the two oscillators are moving through their motions at the same rate. (In Chapter 2, we will learn that these oscillators have the same frequency.) The relative positions of the two vibrating masses cause the difference between the Lissajous figures in Figures 1.8 and 1.9. Notice that in Figure 1.8, the horizontally vibrating mass starts in the center of its path and moves to the right. The same mass in Figure 1.9, however, starts at the right end of its path. (The difference between these two figures will be described in Chapter 2 as a difference in phase. The two masses in Figure 1.9 are vibrating in phase with each other, while those in Figure 1.8 are 90° out-of-phase with each other.)

Figure 1.10 shows a "more complicated" pattern. This pattern still is relatively simple, however, because one mass is oscillating at exactly twice the rate of the other one. We can see

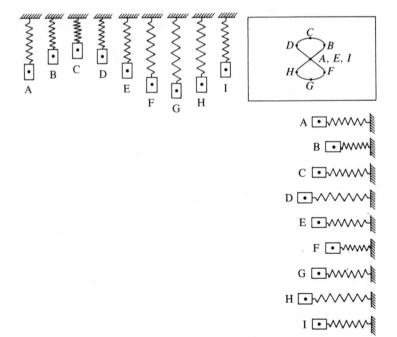

Figure 1.10. The production of a Lissajous figure with two vibrating systems. The horizontally moving mass is moving at twice the rate of the vertically moving mass.

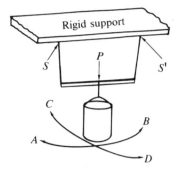

Figure 1.11. A mechanical device to produce Lissajous figures

easily that the rate of one oscillator is twice that of the other by tracing the dot around the figure-eight path. As the dot moves up and down once (starting and ending at A), it moves back and forth twice. Notice also that in the pictures of the oscillating masses, the one moving horizontally moves through two complete cycles during the time the vertically moving mass completes one cycle. Whenever one mass oscillates at a rate that is a ratio of small numbers (such as ⅔, ¾, 1, ⅓, ³⁄₂, 2, or 3) times the rate of the other oscillator, the pattern will be a relatively simple one. This happens because the dot on the screen returns to its starting point fairly quickly; thus, the dot retraces the same path rather than creating new ones.

Other devices also can be used to produce both types of curves we have seen on an oscilloscope. Let me describe one such device. We make a pendulum by using string to suspend a container for sand (an empty tin can, for example) from a bar, which in turn is hung from a rigid support (Fig. 1.11). We punch a small hole in the bottom of the container so that when we fill it, the sand will run out in a slow, steady stream. Now, as indicated in the figure, the pendulum can swing two different ways: parallel to the bar (path A–B) and perpendicular to the bar (path C–D). The two strings holding the bar prevent it from moving when the container swings along path A–B; thus, the moving container is a pendulum that is swinging from point P. In contrast, when the container moves along C–D, the bar moves also, and the container is a pendulum swinging from the supports at S and S'. Consequently, we have a short pendulum when the container swings along A–B and a long pendulum when the container swings along C–D. As we will see in the next chapter, the rate at which a pendulum swings depends on its length; short pendula swing faster than long ones. Thus, we have two vibrations with different rates, which are moving in perpendicular directions. This is exactly what we had on the oscilloscope when we observed Lissajous figures, and we can observe the same patterns with this mechanical device. The container will swing in both directions simultaneously when we release it from any position between A and D, and the trail of sand falling out of the container onto the surface will form a Lissajous figure.

The oscilloscope is a simpler device to use for observing Lissajous figures and vibration patterns because we can use electronic devices to generate the vibrations. With such generators, it is much easier to control the rate of vibration; thus, different figures are easier to create.

Both of the oscilloscope displays I have discussed are useful for describing and understanding vibrations. Of necessity, Lissajous figures can be used only when two vibrations are being compared. However, one of the two vibrations could be a standard. In such a case, a single vibration can be analyzed. The displacement versus time pattern has much wider usefulness, and we will be using such patterns often in our discussions of vibrations.

An interesting display of vibrations can be developed using two oscilloscopes and a stereo system. Since the two channels of a stereo system are independent, they serve nicely as two sources of vibrations for Lissajous figures on one of the oscilloscopes. On the other oscilloscope, we can display the vibration pattern for either of the stereo's channels. Such a demonstration requires, of course, some expensive equipment.

Summary

Physicists use the concept of waves to describe and understand a wide range of things that occur in nature. From these phenomena, we will look at two in detail: sound and light. Since neither sound nor light waves can be seen directly, we will gain an understanding of wave behavior by observing mechanical waves on strings and on the surface of water.

Waves are the movement of vibrations from one place to another. Thus, we will study vibrations first. We again will start our study with mechanical vibrating systems (or oscillators) that we can observe directly. Mechanical vibrations are the source of all sounds, and we can use an oscilloscope to help us observe and study such vibrations.

We use oscilloscopes in two ways. With the first method, we get a picture called a *vibration pattern,* which shows how the position of the vibrating object changes with time. The second picture, a Lissajous figure, occurs when we use the oscilloscope to compare two different vibrations. A vibration pattern of a sound is narrower for higher pitches and taller for louder sounds.

Review Questions

The first four questions are based on the four vibration patterns shown in Figure 1.12, which shows sounds observed with an

Figure 1.12. For use with review questions 1–4

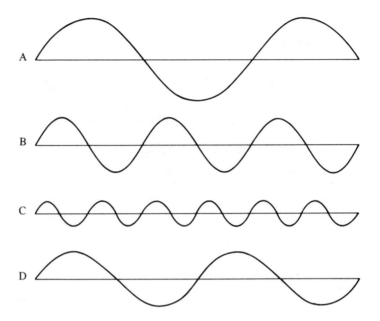

oscilloscope under the same conditions (that is, with the same settings).

1. Which sound has the highest pitch?

2. Which sound has the lowest pitch?

3. Which sound is the loudest?

4. Which sound is the softest?

5. Suppose the planet Mars blows up. About how long will it take before we hear this tremendous explosion?

    a. A few seconds

    b. A few hours

    c. A few days

    d. A few weeks

    e. We never will hear the explosion.

6. For the Lissajous figure shown in Figure 1.13, the rate of oscillation in the vertical direction is _____ times the rate of oscillation in the horizontal direction.

    a. two

    b. three

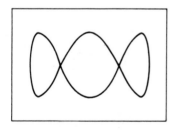

Figure 1.13. For use with review question 6

c. four

d. one-third

e. one-fourth

2

Simple Vibrations

Important Concepts

- A simple oscillator is described using the following terms:

  1. *Equilibrium position*—the location of the vibrating object when the object comes to rest.

  2. *Amplitude*—the maximum displacement of the vibrating object from its equilibrium position.

  3. *Frequency*—the number of oscillations occurring during each unit of time (usually per second).

  4. *Period*—the time required for one cycle of the motion.

  5. *Damping Time*—the time required for the amplitude to change from any value to one-half that value.

- The values of the frequency and period are always reciprocals: $f = 1/T$ and $T = 1/f$.

- The values of the frequency, period, and damping time are inherent properties of an oscillator.

- The motion of an oscillator can be displayed in a displacement versus time graph. For a simple oscillator, this graph is a sine curve.

- The amplitude, period, and frequency of an oscillator can be read from a displacement versus time graph.

- The term *phase* (or *phase angle*) describes the location of an oscillator within a cycle.

- One way to start an oscillator vibrating is by driving it with another vibration. The resulting amplitude shows how well the oscillator responds to the driving vibration.

- A response curve describes concisely how well an oscillator responds at many different driving frequencies.

- When an oscillator has a large response to a driver, resonance is occurring.

- The response curve for a simple oscillator has only one peak.

- Peaks on response curves are tall and narrow for vibrations with long damping times (that is, lightly damped vibrations), and broad, flat peaks occur for heavily damped vibrations.

- When peaks on the response curves of two oscillators have overlapping regions, the oscillators will resonate.

- Adding two independent vibrations to produce a third is called superposition. If the two original vibrations have almost the same frequencies, the resulting vibration will exhibit a varying amplitude; we call this behavior *beats*.

The word *vibration* is used to describe one of the many possible motions that occur in the world around us. Vibrations can be desirable. The vibration of the pendulum in a grandfather clock and the balance wheel in a wristwatch, for example, are bases for good timekeeping. On the other hand, vibrations such as a rattle in a car are annoying or dangerous. There are even cases on record of how vibratory motion has produced disastrous effects: a bridge across the Tacoma Narrows in the state of Washington was destroyed by large vibrations in 1940.

Description of Vibratory Motion

Even though vibrations occur in many different ways, a common vocabulary describes them. We will develop our understanding of this vocabulary by looking first at simple vibrating mechanical systems. Figure 2.1 and 2.2 show the way two simple systems vibrate. A simple pendulum like that in Figure 2.1 can be constructed by hanging a weight (called the pendulum *bob*) on a light string attached to a support. The system in Figure 2.2 consists of a mass or weight hung on the end of a coil spring that is attached at its other end to a support. (This setup is, of course, the oscillator we used in Chapter 1 to describe the patterns on an oscilloscope screen.)

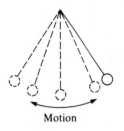

Motion

Figure 2.1. A simple pendulum

When we observe either system, we first note their motions are repetitive. The pendulum swings back and forth over and over again, and the mass on the spring bobs up and down over and over again. We term such motion *periodic* or *cyclic,* and we call each repetition of the motion a *cycle*. Note that we can decide to start the description of a cycle anywhere in the motion; the cycle then finishes when the object returns to the same point. For example, Figure 2.1 shows the pendulum motion starting at the right-hand end. When the pendulum has swung all the way over to the left-hand end, it has completed one-half of a cycle. When it swings back all the way to the right again, it has completed one full cycle.

If we work with either the pendulum of Figure 2.1 or the spring-mass system of Figure 2.2, and if we wait awhile, the system will come to rest. In its resting position, the pendulum hangs straight down (which is the midpoint of its motion), and the mass on the end of the spring rests in the middle of where it was moving. In each case, this position is called the *equilibrium position* of the oscillator. We describe the location of the oscillating object (the pendulum bob, for example) relative to this equilibrium position. Thus, with a simple oscillator, the vibrating object moves along a path that extends equally on each side of the equilibrium position. The maximum distance that the object moves away from the equilibrium position while vibrating is termed the *amplitude,* and the total length of the object's motion is twice the amplitude.

We use two different terms to describe how rapidly an oscillator moves through its periodic motion. The first term, the *period,* is defined as the time taken to complete one cycle of motion. As an example, consider a pendulum that is made from a weight suspended by a string about one meter long. The size of the weight is not important; the weight just must be heavy enough to pull the string taut. Due to the string's length, this pendulum will have a period of about 2 seconds (secs). We will use the symbol T to represent the period, so $T = 2$ secs for this oscillator. The other way to specify how rapidly an oscillator completes its cyclic motion is with the term *frequency*. Frequency is defined as the number of cycles that are completed in one unit of time (the

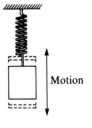

Motion

Figure 2.2 A spring-mass system

unit of time could be, for example, a second, a minute, an hour, a day, or a year). We will use the symbol f to stand for the frequency. In the example of the pendulum with a one meter string, f is about $\frac{1}{2}$ cycle per sec. The unit of cycles per second is given a special name: *hertz* (*Hz*). (This is named for Heinrich Hertz, a German physicist who first produced electromagnetic waves.) Thus, 120 Hz means 120 cycles per sec, and for the pendulum described previously, $f = \frac{1}{2}$ Hz.

Due to the way they are defined, frequency and period will always have values that are reciprocals of each other for a given oscillator: $T = 1/f$ or $f = 1/T$. We saw this reciprocity in the example just cited where $T = 2$ sec and $f = \frac{1}{2}$ Hz.

Depending upon the system being observed, measurement of the frequency and/or the period can be done in a number of different ways. For a typical pendulum and mass on a spring, such as those in Figures 2.1 and 2.2, the motion is usually slow enough to allow us to count the number of cycles that occur in a given amount of time. For example, if we want to determine the period of a pendulum, we start the pendulum swinging and use a stopwatch to measure the time required for one complete cycle. To achieve more accuracy, we might actually measure the time required for ten cycles and then divide that time by ten to get the time for one cycle. In a similar way, we can measure the frequency. Since a typical pendulum usually executes very few cycles in one second, we can achieve a more accurate frequency determination by counting the number of cycles during a longer time interval. The frequency then would be the cycle count divided by the number of seconds in that time interval. Referring again to the pendulum with a one-meter string, we measured the time for twenty cycles in one experiment; the time was 39.8 sec. Thus, the time for one cycle (the period) for that oscillator was 1.99 sec (39.8 sec ÷ 20 cycles = 1.99 sec per cycle). Using the same pendulum, we counted fifteen oscillations in a 30-second interval, so the frequency was $f = \frac{1}{2} = 0.5$ Hz.

Experimental Uncertainty (A Digression)

When the results of frequency- and period-determining experiments on an oscillator are multiplied, the product obtained from the values should be one. It is likely, however, that although the result is close to one, it is not exactly one. This result occurs even when no mistakes were made during measurements, and it illustrates an important fact inherent in any experiment: Whenever a measurement is performed, the results obtained always have some *uncertainty*. Such experimental uncertainty can be reduced by refining the techniques used to obtain results, but uncertainty never can be eliminated entirely. In the experiment just described, for example, the use of an automatic timing device that would record time intervals accurately to one one-thousandth of a second probably would make the product of the period and frequency nearer to one. However, even with all the technology available today (or with any technology imaginable), the result still would have some uncertainty.

The reason for discussing experimental uncertainty is two-fold. First, it is important to recognize that such uncertainty exists for every scientific experiment. The second reason is more immediate: It explains why the product of a measured frequency and a measured period is not necessarily one.

Sine Curve Readings

We saw in Chapter 1 that when we display the motion of a simple oscillator on an oscilloscope, we see a sine curve. We can read the values of the amplitude, frequency, and period from a properly labeled sine curve. Remember that these curves display information about how the position of the oscillator (that is, its displacement from the equilibrium position) changes as the time changes. Therefore, a "properly labeled" sine curve must contain information about the size of the displacement (which is plotted in the vertical direction) and the time (which changes along the horizontal direction). We call this size information the scale of the curve.

Figure 2.3 shows a sine curve with all the necessary information provided. In the vertical direction, the displacement is given in centimeters (cm), and the horizontal time scale is in seconds. The straight line through the center of the curve is the equilibrium position (the displacement is zero), and where the sine curve is above this line, the oscillating object is above the equilibrium position. Since the top of the sine curve is opposite the 6 cm mark (and its bottom is opposite the −6 cm mark), the amplitude of this oscillation is 6 cm. Now, here is how we dis-

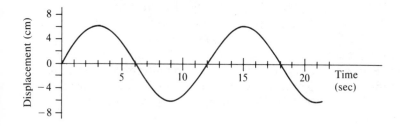

Figure 2.3.   A sine curve showing the displacement of an oscillator versus time

cover that the time required for one complete cycle is 12 secs. As noted earlier, we can start the cycle at any point; we will look at two possible starting points. In the first case, we start where the oscillator is at the top of its motion (this happens at the 3-sec mark). The next time the oscillator reaches the top occurs at the 15-sec mark—12 sec have elapsed. The second starting point we consider is at the beginning of the curve (when the time is zero). At this time, the vibrating object is at its equilibrium position moving upwards. At the 6-sec mark, the oscillator is again at its equilibrium position, but it is moving downward now, so the cycle has not been completed. At the 12-sec mark, the cycle is complete because the object is moving upwards through the equilibrium position again. Since the time for one cycle is 12 sec, the period and frequency of this oscillation are, respectively, $T = 12$ sec and $f = \frac{1}{12}$ Hz.

The sine curve in Figure 2.4 represents another oscillation, and we can use the same techniques to find the values of this oscillation's amplitude, period, and frequency. It should be clear to you that the amplitude is 2.5 cm, the period is 2 sec, and the frequency is $\frac{1}{2}$ Hz. One other notable difference exists between the curves in Figures 2.3 and 2.4: the way the curves start. In Figure 2.3, the curve starts at the equilibrium position; whereas, the curve in Figure 2.4 starts above the equilibrium position. We now will discuss how we describe this difference.

Figure 2.4.   A second sine curve

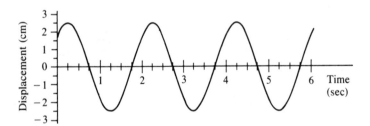

Phase of Vibratory Motion

Another describable aspect of vibrating systems is the location of the moving object within its cycle. In order to describe this, we make use of a similarity between oscillatory motion and circular motion. We have seen how the motion of an oscillator is repetitive, one cycle following another. The same thing can be said about a point moving around the circumference of a circle. After the moving point has made one complete revolution, the point starts all over again. We can specify the location of the point on the circle by using the angle between a line from the center of the circle to the point and another line from the center of the circle to the starting point. We show this in Figure 2.5. The point starts at A and moves counterclockwise around the circle. When it reaches B, we say the point has moved through an angle of 90°. When the point reaches C, it has moved through a total angle of 180°, and at D, the point has moved through a total angle of 270°. Finally, when the moving point is again at A, it has moved through an angle of 360°. But at A, the point also is starting a new cycle, so we start the angle measurement over again and say the point is at an angle of 0° as well.

We use the same circular motion idea to describe the location of a vibrating object within a cycle. The term used to describe this motion through the cycle for an oscillator is *phase*, or *phase angle*. As an example, consider the pendulum in Figure 2.6. We have arbitrarily chosen the starting point for this cycle at the right-hand end of the swing as shown in picture A. During a full cycle of its motion, the pendulum will swing all the way over to the left and back all the way to the right, and the phase angle will change from 0° to 360°. As shown in the figure, when the pendulum moves from the starting point at the right end (in picture A) to the equilibrium position (in picture B), the pendulum has completed one quarter of a cycle, so the phase angle has changed from 0° to 90°. Then, as the pendulum moves to the left end of its swing (in picture C), the phase angle increases to 180°. This action is the same as the point on the circle moving halfway around its circular path. As the pendulum completes its cycle, it first moves back to the equilibrium position where the phase angle is 270° and then back to the starting position at the right-hand end where the phase (or phase angle) is 360° (or 0° for the next cycle).

We can, of course, choose the beginning of the cycle for an oscillator at any point of its motion. We illustrate this in Figure 2.7 where two more sequences of pendulum pictures are shown, and each sequence starts its cycle at a different location. Now, we have a way of describing the different starting points of the sine curves in Figures 2.3 and 2.4: When they start, the sine curves have different phases. Unlike a pendulum having no position that must be given a phase angle of 0°, the mathematical basis of a sine curve defines its phase. The curve in Figure 2.3 starts with a zero phase angle while the curve in Figure 2.4 has a zero phase angle at the 1.75-sec mark.

Because we can start a cycle of an oscillator at any point, talking about phase changes is often more useful. Note that no

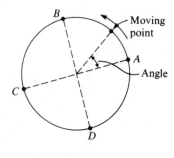

Figure 2.5. A point moving around a circle and the angle used to specify the location of the point

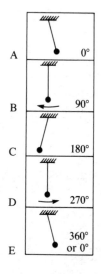

Figure 2.6. A pendulum moving through its cyclic motion

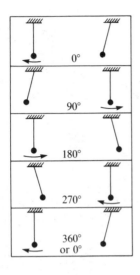

Figure 2.7. Two pendula moving through their cyclic motion. The two pendula are 90° out of phase with each other.

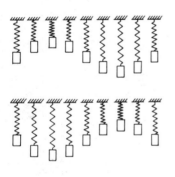

Figure 2.8. Two spring-mass systems oscillating in antiphase with each other

matter where we start the cycle for the pendulum, the phase always changes by 180° as the pendulum moves from one end of its swing to the other end. The phase angle also changes by 180° as the pendulum moves from the equilibrium position out to one end and back to the equilibrium position again. Finally, the phase always changes by 90° when the pendulum moves from the equilibrium position to one end or from one end to the equilibrium position.

We also use the phase concept to compare the movement of two oscillators such as two pendulua. When the pendula are swinging together (that is, moving in the same direction at the same time), we say they are *in phase* with each other. On the other hand, if the two pendula are moving exactly opposite to one another, we say they are 180° out-of-phase with each other, or they are in *antiphase* with each other. Of course, this concept of phase can be applied to any oscillator. Figure 2.8 shows two spring-mass systems that are vibrating in antiphase with each other.

In Chapter 1, we also discussed Lissajous figures, which are produced by the movement of two oscillators. In particular, we saw in Figures 1.8 and 1.9 that two oscillators with the same frequency can produce different Lissajous figures. The difference between these two cases is the relative phase between the two oscillators. The circular Lissajous figure shown in Figure 1.8 occurs because the two oscillators creating it have a phase difference of 90°. The straight line of Figure 1.9 occurs because the two oscillators are vibrating in phase with each other. When the phase difference between the two oscillators is between 0° and 90°, the resulting Lissajous figure is an ellipse. The ellipse changes its eccentricity as the phase difference between the two oscillations changes. Long thin ellipses result from phase differences that are nearly zero, and almost circular ellipses appear for phase differences near 90°. Thus, Lissajous figures are useful for comparing both the frequencies and phases of two oscillations.

Damping

In this chapter, we are describing oscillators that we call simple. One of the most important characteristics of a simple oscillator is that when the system is set in motion and allowed to move by itself, the oscillator vibrates at only one frequency. Of course, if we change the vibrating system (changing the length of the pendulum, for example), then the frequency will be different. However, for a given oscillator, the frequency always will be the same no matter what the amplitude is. We call this frequency the *natural frequency,* or *characteristic frequency,* of the oscillator.

On the other hand, amplitude can take on many different values. Indeed, when an oscillator is allowed to vibrate freely, its amplitude gradually will decrease because frictional forces always rob the oscillator of energy—and the amount of energy in the oscillator is proportional to the square of the oscillator's amplitude. For some systems, this energy decay or damping occurs very rapidly, while for other systems it occurs much more slowly. We use the term *damping time* to describe this effect quantitatively. Damping time is the time taken for the amplitude of an oscillator to change from any value to one-half of that value.

Let's illustrate this idea of damping time with an example. We will use a pendulum made from a solid metal weight hung on a string about one meter long, and we will describe the displacement of the pendulum in terms of the horizontal distance the weight moves from its equilibrium position. Now, we start the pendulum swinging with an amplitude of 12 cm and measure the time needed for the amplitude to decrease to 6 cm. This time is the damping time. The interesting thing about the damping time is that its value is independent of the original amplitude. Thus, the same amount of time will be required for the amplitude to decrease from 6 cm to 3 cm, as was required for the decrease from 12 to 6 cm. To reiterate: No matter what amplitude we start with, it will take the same amount of time for the amplitude to decay to one-half its starting value; this time is the damping time. Note also that if we wait two damping times, the amplitude decreases to one-fourth of its original value. An oscilloscope trace of a damped oscillator is shown in Figure 2.9.

Resonance

Up to now, we have not worried about how we have started an oscillator moving. For either a pendulum or a spring-mass system, the easiest way to start the motion is to pull the weight away from the equilibrium position and then release the weight. Once released, the weight moves back and forth (or up and down) with a continually decreasing amplitude. The rate at which the

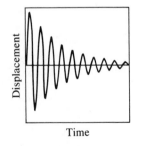

Figure 2.9. The trace on an oscilloscope screen of a damped oscillator

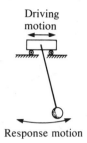

Driving
motion

Response motion

Figure 2.10. A pendulum with a movable mounting

amplitude decreases, of course, is described by the damping time.

Another way to start an oscillator vibrating has important applications in our study of vibrations and waves. As an introduction to this new idea, consider the specially mounted pendulum shown in Figure 2.10. The mounting of this pendulum is special because the string is not tied at the top to a rigid post but, instead, is tied to a post on a cart which can be moved back and forth. Thus, the mounting in this figure can itself vibrate. If we now move this mounting back and forth with a small amplitude (say 2 cm) and at a very low frequency, the bob of the pendulum would also oscillate with the same frequency and an amplitude of about 2 cm. This experiment is very easy to perform yourself. You can easily construct the weight on a string, and you can hold the upper end of the string in your fingers. Moving your hand back and forth serves the same purpose as moving the cart just described. And, as mentioned for the cart, a very low frequency motion of your hand will cause the weight to move back and forth with the same amplitude.

A very interesting thing happens, however, when we increase the oscillation frequency of the mounting (that is, the cart or your hand). The amplitude of the pendulum bob's oscillation will not stay at about 2 cm. Instead, when the mounting's vibration frequency becomes nearly the same as the natural frequency of the free pendulum, the amplitude of the pendulum's vibration will increase dramatically. The pendulum described earlier (with a length of about 1 meter) has a natural frequency of about ½ Hz (period is about 2 sec). Thus, as we move the cart (or you move your hand) back and forth with a period of 2 sec, the amplitude of the pendulum's oscillation will get quite large. As the mounting's vibration frequency increases further, it becomes much larger than the natural frequency of the pendulum, and we see a decrease of the pendulum's vibration amplitude. Finally, at very high vibration frequencies of the mounting, the pendulum's amplitude approaches zero. Thus, at these high frequencies, the mounting oscillates very rapidly, but the pendulum bob barely moves.

This experiment is described technically in the following way. By vibrating the mounting of the pendulum, we are *driving* the oscillator (that is, the pendulum), and when the oscillator starts vibrating, it is *responding*. (Note that in this use, the term *driving* means pushing or forcing.) The frequency at which the mounting vibrates is called the *driving frequency,* and the pendulum's response increases dramatically when the driving frequency is nearly equal to the pendulum's natural frequency. When an oscillator gives a large response such as this, the oscillator is *resonating*.

A child on a swing serves as another good example of resonance. As long as the child "pumps" the swing at just the right time, the amplitude of the swinging motion will increase. In the terms we are using here, the child on the swing forms a pendulum that has a natural frequency. This pendulum is driven by the child pumping the swing, and the rate at which he pumps is the driving frequency. When the driving frequency matches the natural frequency, the amplitude increases, and we say there is resonance. It is also important to recognize that when resonance exists for an oscillator, a large amount of energy is being transferred into the oscillator.

We can use a graph (called a *response curve*) to describe how an oscillator responds when it is driven. A typical response curve is shown in Figure 2.11. The horizontal axis plots the frequency of the driver, and the height of the curve shows the resulting amplitude of the oscillator (that is, the oscillator's response) at each frequency. The peak of the response curve occurs at that particular driving frequency which is equal to the oscillator's natural frequency. Since the response curve in this figure has only one peak, we know the curve describes an oscillator that has only one natural frequency. In this figure, the driving frequency increases as we move out along the horizontal axis. The vertical, dashed line in the middle of the peak (labeled f_R) is located at the natural or *resonant* frequency of this oscillator. Resonance—that is, transfer of energy into this oscillator—will occur for any driving frequency where the response curve is raised above the horizontal axis. The resonance gets stronger as the curve gets higher.

In addition to describing the natural frequency of a vibration, a response curve also gives information about the vibra-

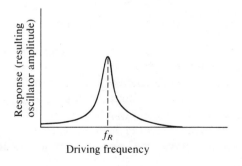

Figure 2.11. A response curve for an oscillator

tion's damping time. This information is contained in the shape of the peak associated with the vibration. When a peak on a response curve is narrow and sharp, the vibration is lightly damped; it has a long damping time. In contrast, a broad, flat peak means the damping time is short; the vibration is heavily damped. The two response curves shown in Figure 2.12, labeled A and B, describe two different simple oscillators. The curves are drawn to the same scales so the shapes and locations of the peaks can be compared easily. Since the center of peak B is located farther to the right than that of peak A, the natural frequency of B is higher than that of A. The natural frequencies of each oscillator are marked and labeled f_A and f_B. The damping time of vibration A is longer than that of vibration B because peak A is narrower.

Thus, a response curve is a very concise way of describing two main features of a vibration. In this chapter, we are looking only at simple oscillators which vibrate at only one natural frequency; the response curve for any simple oscillator has only one peak. More complicated vibrating systems or oscillators, which are considered in the following chapters, may have several different natural frequencies. A response curve for a complicated system will contain as many peaks as the system has natural frequencies, and each peak will give information about the natural frequency and damping time of that vibration.

Figure 2.12. Response curves for two different oscillators. Oscillator A has a smaller frequency and a longer damping time than oscillator B.

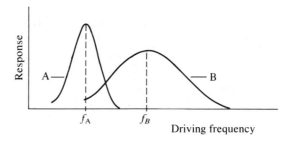

Coupling

Very often, we want to transfer vibrations from one object to another. Sometimes, this transfer can be handled easily, but in other cases, it is more difficult. When the transfer occurs easily, we say good coupling exists between the objects.

To illustrate the concept of coupling, let's consider two examples. In the first example, suspend a standard coat hanger by two threads as shown in Figure 2.13. When swung against a hard object, the coat hanger emits a "clink," one short noise that dies out almost immediately. However, when each thread is wrapped around a finger on each hand, those fingers are inserted into the ears, and the hanger again is hit against a solid object, a truly amazing thing happens: One hears a bell-like clanging emanating from the coat hanger. This new sound was there all the time. It just could not be heard in the first case because poor coupling existed between the coat hanger and the ears; the vibrations had to pass through the air. In the second case, the vibrations were carried from the coat hanger to the ears by passing up the threads and through the fingers; the coupling was much better!

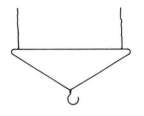

Figure 2.13. A coat hanger suspended by two threads

Another example of coupling occurs with a tuning fork. Before discussing this coupling example, let me first describe the tuning fork. It is a simple oscillator (see Figure 2.14), and we can make it vibrate by hitting it against something or striking it with a hammer or mallet. The natural frequency of a tuning fork depends on its size and shape; generally, longer tuning forks have lower frequencies. We also can lower the natural frequency of a given tuning fork by adding weight to one (or both) of the fork's tines.

The typical laboratory tuning fork does not produce a very loud sound, however. The reason it does not is because vibrations are not transferred easily from the tuning fork to the air surrounding it. This poor coupling is due to the small vibrating surface on the tuning fork. The coupling improves greatly when the base of the tuning fork is held against a large surface that can vibrate. The back of an acoustic guitar or a violin is an excellent surface to use. Sometimes, a table will work well. In order to be a good coupler, a surface must meet one requirement: The surface must respond well to the driving of the tuning fork. Some tuning forks are mounted on boxes that have an open end. These boxes provide good coupling to the air because they have a large vibrating surface.

Other Examples of Resonance

Very interesting examples of resonance occur when we use two vibrating systems. For example, consider two pendula which have equal lengths and are suspended from a common support. If we start one pendulum swinging, then its vibrations will drive the other pendulum into a swinging motion. The amplitude of the second pendulum's motion will increase while that of the first pendulum will decrease. The first pendulum will stop mov-

Figure 2.14. A tuning fork

ing, and then, it will be restarted by the vibration of the second pendulum. This process (where the energy of vibration is transferred back and forth between two pendula) will continue—but with damping—until both pendula are at rest. These two pendula "resonate with each other" because they transfer energy so easily. This transfer happens, however, only when the natural frequencies of the two systems (the pendula) are the same. When only one pendulum is lengthened (or shortened), vibrations (or energy) will no longer transfer.

A similar situation occurs when two tuning forks mounted on boxes are used instead of pendula. By facing the two open ends of the boxes toward each other, we can get good coupling between them. Now, if the frequencies of the tuning forks are the same, then the two systems will resonate. Like the pendula demonstration, we start one of the tuning forks vibrating by hitting it with a hammer. Unlike the pendula experiment, however, we cannot see which fork is vibrating. The only observable effect is the sound of the vibration, and it is difficult to detect which fork is vibrating from the sound alone. Thus, after hitting one fork (which we will call number one), we stop its vibration by touching it. The sound persists, however, because fork number two is vibrating now. (Again, this does not happen if the two tuning forks have different frequencies.)

As mentioned, some difference may exist between the natural frequencies of the two vibrating systems. But as long as the response curves of the systems have a common non-zero region, we will get resonance. In other words, at least some part of a peak on each response curve must occur at the same frequency.

A pair of tuning forks that are mounted on acoustic resonance boxes. A mallet for striking the forks also is shown.

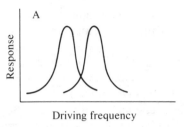

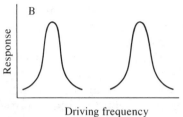

Figure 2.15. Two sets of response curves. The oscillators in A will resonate with each other; those in B will not.

Figure 2.15 shows response curves for two sets of systems. The overlapping response curves shown in A reveal the two systems will resonate; whereas, the response curves shown in B signify no resonance will occur.

Measurement of Rapid Vibrations

Thus far, we have discussed the measurement of vibrations that occur slowly enough for us to observe with the naked eye. However, we also must study other oscillators that move too rapidly to observe with the naked eye. One such oscillator is the tuning fork described previously. Typically, a tuning fork has a frequency that is greater than 100 Hz. In order to observe and measure such a vibration, we need to use some type of special instrument. Even without the use of special equipment, however, we can realize the tuning fork vibrates. If you strike the tuning fork against a hard surface, the tuning fork will start vibrating, and it will emit a tone of constant pitch. By placing the top end of the tuning fork tine lightly against your cheek, you can feel the tine's motion. Of course, the tuning fork is moving much too fast to count the number of vibrations that occur in a second. We need an instrument to do the counting for us or to slow down the motion so we can count the vibrations ourselves.

As described in Chapter 1, we can use the oscilloscope to "see" the vibration of the tuning fork. The easiest way to see the vibrations is to connect a microphone to the oscilloscope in order to detect the sound created by the vibrating tuning fork. The vibration pattern on the oscilloscope screen will appear similar to that shown in Figure 2.16. As we described in Chapter 1, Figure 2.16 is also a sine curve. Recalling the oscilloscope discussion from Chapter 1, we can note the tops and bottoms of the sine curve occur when the tines of the tuning fork are at the ends of their motion. Thus, for example, each time the curve reaches the top, the tines might be closest together, and then, each time the curve reaches the bottom, the tines might be farthest apart. (The

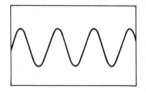

Figure 2.16. The trace on an oscilloscope of the vibrations of a tuning fork

relationship between the curve and the position of the tines could be the other way around, but this is not important for our discussion.)

Using the techniques described earlier, we easily can find the frequency and period of the fork's vibration from the sine curve. (The oscilloscope supplies the scale for the horizontal time axis.) Determining the amplitude of the fork's vibration is more difficult, however, because the oscilloscope is not connected directly to the fork; the oscilloscope detects the sound created by the fork. The height of the sine curve does depend on the fork's amplitude, so we can tell when the amplitude is larger or smaller. However, the height of the curve also depends on how close the microphone is to the tuning fork. When we hold the tuning fork close to the microphone, the curve is taller than when we move the tuning fork farther away. We find a similar effect when we consider the loudness of the tuning fork. It is louder when we hold it closer to our ear.

Another device we can use to observe and measure rapid vibrations is called a *strobe light,* or *strobe.* Using this light, we can make a vibrating or rotating object appear to slow down or even stop. This "slowing" occurs because the light flashes on for a brief instant of time (and then off) at regular time intervals. The time interval between flashes (or the rate at which flashes occur) can be adjusted. When we adjust the rate of the light flash so that the rate equals the frequency of the oscillator we are observing, then the oscillator will appear to stand still. In addition, if the flash rate of the light is slightly different from the frequency of the oscillator, then the oscillator will appear to move slowly through its motion. As an example of how this works, let's consider our old friend, the pendulum. Imagine that we have a pendulum swinging in a dark room, and that someone turns the light on and off very rapidly each time the pendulum is at the right end of its swing. The only thing we could see would be the pendulum when it is at the right end of its swing; it would appear to be standing still at that point. In this case, the frequency of the light source and the frequency of the pendulum are equal. When the person working the light switch is a little slow the pendulum will move a little farther through its cycle during each dark period. In this case, we will observe the pendulum swinging very slowly rather than standing still.

Since it has an adjustable calibrated flash rate, the strobe light can be used to measure the frequency of vibrating objects. A complication exists, however, and we can see it by again referring to the pendulum. Suppose the pendulum makes two complete swings between flashes of light. In this case, the pendulum still appears to be standing still; however, the frequency of the light flashes is simply one-half the frequency of the pendulum. The pendulum also would appear to stand still if the frequency of the light source were one-third that of the pendulum (or one-fourth, or one-fifth, etc.). On the other hand, consider what would happen if the frequency of the light source were twice that of the pendulum. Then, the light would flash on twice during each cycle of the pendulum. So, if the light flashes on when the pendulum is at the extreme right end of its motion, the light also would flash on when the pendulum was at the extreme left end of its motion. In this situation, we would observe two stationary objects which look like the pendulum, one at each end of the pendulum's motion. We also would observe multiple stationary images of the pendulum for those cases where the frequency of the light is any integral multiple of the frequency of the pendulum (that is, two times, or three times, or four times, etc.). Yet, another complication exists: Some of the multiple images may be concealed because they lie on top of one another. An example of this complication occurs when the frequency of the light is twice that of the pendulum, but the light is illuminating the pendulum at its equilibrium position. Although illuminated twice during each cycle, the pendulum is located at the same position each time. And since the light flashes on and off so quickly, we cannot tell which way the pendulum is moving.

In summary, the strobe can be used to measure the frequency of an oscillator, but the strobe must be used carefully. To do this measurement with the strobe light in practice, adjust the strobe's frequency until a single image of the oscillator is obtained. Then, double the strobe's frequency to see if another image is produced. If the frequencies of the light and oscillator were originally identical, a second image of the oscillator should appear when the strobe light's frequency is doubled.

We also can make rotary motion appear to slow down or stop using the strobe light because rotary—like vibratory—motion is cyclic. When using a strobe light to measure the frequency of a rotating object, the same problems can occur as when measuring a vibrating object's frequency. Thus, stopping the rotary motion with the strobe light does not guarantee that the frequencies are matched. For an interesting example, consider a four-bladed fan that has three blades painted white and the fourth blade painted red. When the fan is running and illuminated with the strobe light at the same frequency, the fan will appear stationary with one red blade and three white blades. Now, if the frequency of the strobe light is doubled, there will be two white blades and two pink blades. This illusion happens because there are overlapping double images. The appearance of each pink blade results from the presence of two images: a red blade and a white blade. The eye perceives the mixture of red and white as pink. The appearance of each white blade also results from the presence of two images: two white blades.

The motion of a vibrating tuning fork can be slowed down in the same way as the motions of the pendulum and fan described above. When slowed down, the ends of the tuning fork tines can be seen oscillating as they move alternately toward and away from each other as shown in Figure 2.17.

Other Vibrations

An electronic instrument called a signal generator produces electrical vibrations. These electrical vibrations can be viewed on an oscilloscope screen and used to power a loudspeaker to create sound vibrations. As a source of vibrations, the signal generator has a number of advantages over the other sources we have discussed thus far. The primary advantage is that both the frequency and amplitude of the vibration can be adjusted easily. Also, the vibration produced by the signal generator does not decay; the vibration has an infinite damping time. Another feature of some signal generators is the ability to produce vibrations which show different patterns on the oscilloscope screen. Figure 2.18 illustrates some possible patterns. The first pattern, of course, is the sine curve which has been discussed. The other three patterns have obvious names: They are triangular, square, and sawtooth vibration patterns. As mentioned in Chapter 1, physicists usually call these vibration patterns waveforms because the vibrations are carried by waves. I am refraining from doing this, however, because we are not introducing the subject of waves until later. Although the latter three vibration patterns may appear to be as simple as the sine curve, they are not. The sine curve has an obvious quality which the other three are lacking: It is smooth everywhere. Each of the other three curves has sharp bends in it, and these bends create the complexity of the pattern.

Superposition and Beats

Figure 2.17. The motion of a tuning fork

A very fundamental idea in physics is called the *principle of superposition*. This principle says that when two (or more) actions produce a change on something, the net change is the sum of the two changes produced by the individual actions. Let's consider a few examples to help illuminate this important principle. For the first example, suppose you are standing up and John is pulling on your right hand while Marsha is pulling on your left hand. If John would move you two meters to the right (when Marsha was not pulling) and Marsha would move you one meter to the left (when John was not pulling), then with both pulling, you would move one meter to the right.

As another example, consider a room that has both a heater and an air conditioner in it. Suppose the room temperature rises 20° when the heater operates alone, and the room cools 15° when the air conditioner operates alone. Then, the principle of superposition says that when both are turned on, the room temperature will rise 5°. If the room had two heaters instead of the heater and air conditioner, the principle would work in the same way. That is, if heater number one, by itself, causes the room temperature to rise 20°, and if heater number two, by itself, causes a 10-degree rise, then both heaters together will produce a 30-degree temperature rise. Thus, when the principle of superposition applies, we just add the individual changes to get the total effect.

Another example—which has something to do with our main subject—of the principle of superposition occurs when two vibrations act at the same time on an object. Such a case occurs when two sound vibrations arrive at an eardrum simultaneously. Each of these vibrations tries to move the eardrum, and the eardrum's resulting movement is the sum of the two individual motions. If the two original vibrations have slightly different frequencies, then the resulting motion is a vibration with a varying amplitude. The varying amplitude for sound vibrations causes a varying loudness that we call *beats* because it can sound like the repetitive beating of a hammer on an anvil. Note, however, that the term beats is not limited to the effect produced with sound; *beats* applies to the regularly varying amplitude that occurs when any two vibrations are added together.

Let's look more closely at beats to see why the amplitude of the resulting vibration changes in a regular way. In Figure 2.19, we see two sine curve vibration patterns of different frequencies and the vibration pattern resulting from their superposition. The upper curve has a frequency of 300 Hz, and the middle curve has a frequency of 320 Hz. At the left side of the diagram (and the right side as well), the two vibrations are in phase with each other. Thus, their two motions add together to produce a bigger

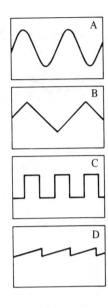

Figure 2.18. Oscilloscope traces of vibrations produced by an electronic signal generator

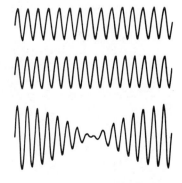

Figure 2.19. The superposition of two sine curves to produce beats

Figure 2.20. The superposition of two sine curves producing beats. The beat pattern is shown over a longer time range.

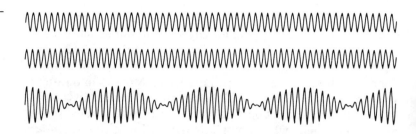

motion as shown in the lower curve. However, in the center of the diagram, the two vibrations are in antiphase with each other, so they cancel. Thus, the resulting amplitude varies from being large at the left side to zero in the center, and then large again at the right side. The regular change in the amplitude of the resulting vibration appears clearly in Figure 2.20, which is the superposition of the same two vibrations drawn with a different scale. The rate at which the amplitude changes is related to the frequencies of the two vibrations being superposed (that is, added together). By counting the number of cycles of each vibration, we can establish two facts. First, the resulting vibration has a frequency that is the average of the two frequencies producing it. Second, the beat frequency (that is, the repetition rate for the varying amplitude) is equal to the difference of the two vibrations' frequencies. We can write the following formulas to express these ideas concisely:

$$f = \frac{1}{2}\,(f_1 + f_2) \tag{2.1}$$

$$f_B = |f_1 - f_2| \tag{2.2}$$

where f_1 and f_2 are the frequencies of the original two vibrations, and f and f_B are the frequencies of the resulting vibration and the beat frequency, respectively. In the example shown in Figures 2.19 and 2.20, the two original frequencies are $f_1 = 300$ Hz and $f_2 = 320$ Hz. Thus, the resulting vibration pattern has a frequency $f = 310$ Hz and a beat frequency $f_B = 20$ Hz.

Summary

In this chapter, we have discussed a large number of ideas associated with simple vibrations. A simple oscillator is an object that produces a simple vibration when it moves. We call the resting position of an oscillator its equilibrium position, and we describe the vibration using the terms amplitude, frequency, period, and damping time. The amplitude is the oscillator's maximum displacement from the equilibrium position; the frequency is the number of vibrations during each second; the period is the time required for one oscillation; and the damping time is the time required for the amplitude of the oscillator to change from any value to one-half of that value. For any oscillator, the magnitudes of its frequency, period, and damping time depend on its construction, while the amplitude is proportional to the energy of the vibration. In all cases, the frequency and period are reciprocals of each other.

The motion of an oscillator can be displayed in a graph of displacement versus time; we call this graph a vibration pattern. The vibration pattern of a simple oscillator is always a sine curve, and when the vibration produces a sound, it is always a pure tone. We saw how to determine the values of the frequency, period, amplitude and damping time from a displacement versus time curve.

We use the concept of phase or phase angle to describe the location of an oscillating object within a cycle. During each cycle of a vibration, the phase changes by 360°. Thus, a phase change of 180° means the vibrating object has moved through one-half a cycle. We also use the concept of phase to relate two vibrations to each other: When two oscillators are vibrating in phase, they are moving together. We describe exactly opposite motion by the two oscillators by saying they are vibrating in antiphase or vibrating 180° out-of-phase.

We can start an oscillator moving by driving it with a vibrating force. The response of the oscillator to this driving force (that is, the resulting amplitude) depends on the relative frequencies of the driver and the oscillator. A response curve shows how the amplitude of the oscillator varies as the driving frequency is changed. Resonance occurs when there is a large transfer of energy from the driver to the oscillator; this transfer happens when the oscillator has a large amplitude (or a large response). The response curve for a simple oscillator has only one peak; it is centered on the natural frequency of the oscillator. The shape of the peak depends on the damping time of the oscillator: A tall, narrow peak means the damping time is long, while a short damping time produces a short, wide peak. When two oscillators have overlapping peaks on their response curves, they will resonate with each other.

The principle of superposition says that when two vibrations act on the same object, the object's displacement will be the sum of the displacements produced by the two vibrations individually. When two vibrations having nearly the same frequencies act together, beats are produced. The phenomenon of beats is just a vibration with a varying amplitude. The rate of amplitude variation is equal to the difference between the two frequencies producing the beats.

Suggested Readings

The first few chapters of *Horns, Strings, and Harmony,* by Arthur H. Benade (Garden City, N.Y.: Anchor Books/Doubleday, 1960) are devoted to the description of vibrations and waves. The treatment is very similar to that employed here in Chapters 2 and 3, but a few additional topics are covered.

An elementary introduction to the physics of sound can be found in the book *Exploring Sound,* by Alexander Efron (New York: Hayden Book Company, 1969). The early chapters of this book also discuss vibrations to a suitable degree.

Review Questions

1. A pendulum swings back and forth, completing 20 cycles in 10 sec. Therefore, its period and frequency are, respectively,

    a.   0.5 sec and 0.5 Hz.

    b.   0.5 sec and 2 Hz.

    c.   2 sec and 2 Hz.

    d.   2 sec and 0.5 Hz.

    e.   10 sec and 20 Hz.

Figure 2.21 shows displacement versus time graphs of several simple oscillators (sine curves). Questions 2 through 6 are based on this figure.

2. The amplitudes of oscillators A and B are, respectively,

    a.   4 sec and 4 sec.

    b.   3 cm and 2 cm.

    c.   6 cm and 4 cm.

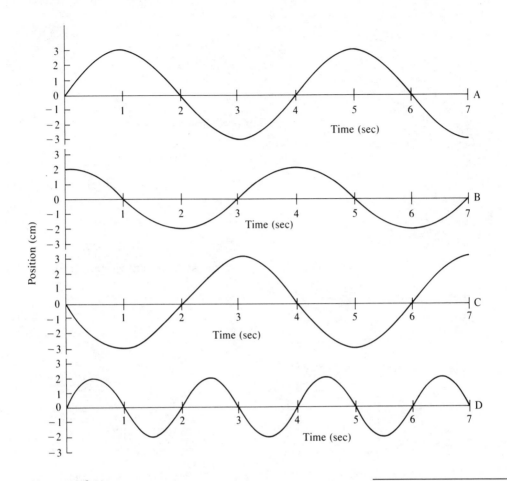

Figure 2.21. For use with review questions 2–6

d. ⅓ cm and ½ cm.

e. ½ cm and ⅓ cm.

3. The two oscillators that differ only by phase are

a. A and B.

b. A and C.

c. A and D.

d. B and C.

e. C and D.

4. The frequency of oscillator A is

a. 3 cm.

b.   4 sec.

c.   2 sec.

d.   ½ Hz.

e.   ¼ Hz.

5.   The period of oscillator D is

a.   2 cm.

b.   1 sec.

c.   2 sec.

d.   1 Hz.

e.   ½ Hz.

6.   The difference in phase between oscillators A and C is

a.   0°.

b.   90°.

c.   180°.

d.   270°.

e.   360°.

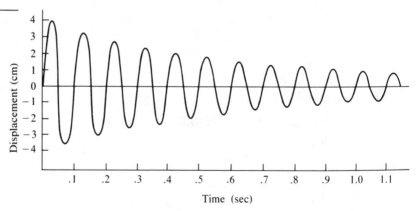

Figure 2.22.   For use with review question 7

7.   The damping time of the oscillator described by the displacement versus time graph of Figure 2.22 is

a.   ¹⁄₁₀ sec.

b.   10 sec.

 c. ½ sec.

 d. 2 sec.

 e. 1 sec.

8. A pendulum has a damping time of 10 sec and swings with an amplitude of 16 cm. How long will it take for its amplitude to reach 4 cm?

 a. 5 sec.

 b. 10 sec.

 c. 15 sec.

 d. 20 sec.

 e. 30 sec.

9. The amplitude of a pendulum drops from 24 cm to 3 cm in 120 sec. Therefore, the damping time of this pendulum is

 a. 20 sec.

 b. 30 sec.

 c. 40 sec.

 d. 60 sec.

 e. 120 sec.

10. A strobe light illuminates a rotating, two-bladed fan. When the flash rate of the strobe is 1600 flashes/min., two stationary blades are seen. Therefore, the fan may be rotating at

 a. only 400 revolutions per minute.

 b. only 800 revolutions per minute.

 c. only 1600 revolutions per minute.

 d. either 400 or 800 revolutions per minute.

 e. either 800 or 1600 revolutions per minute.

11. A fan turns at a speed of 30 revolutions per second. When it is illuminated with a strobe flashing at 30½ flashes per second, we see

 a. a single stationary fan.

 b. two stationary fans that are slightly displaced from one another.

    c. a single fan rotating forward at a slow speed.

    d. a single fan rotating backward at a slow speed.

3

Complicated Vibrating Systems

Important Concepts

- A complicated vibrating system can vibrate in more than one way.

- A system has the same number of normal modes of vibration as it has degrees of freedom. For an elastic string with N beads, there are N normal modes and N degrees of freedom.

- Each normal mode has its own frequency and damping time.

- A system's first normal mode of vibration always has the lowest frequency. It also is called the fundamental mode of vibration.

- The second (third, fourth, etc.) normal modes also are called the first (second, third, etc.) overtones.

- Whenever the higher normal mode frequencies are integral multiples of the fundamental frequency, we also call the normal modes harmonics.

- Any complicated vibration of an extended object (such as a string, a plate, or a block of jello) can be built up using a combination of the object's normal modes.

- At a node or along a nodal line, there is no motion of the extended object as it vibrates. Maximum displacement occurs at antinodes or along antinodal lines.

- The vibration recipe describes the amplitude of each normal mode in a complicated vibration.

- When we drive a string at a particular place, no normal modes that have nodes at that point can be in the vibration recipe.

In this chapter, we will be investigating more complicated systems than we looked at in Chapter 2. The major distinguishing feature we use to classify a vibrating system as simple is that the vibrating system has only one natural frequency of vibration. In contrast, complicated vibrating systems have more than one natural frequency. The actual number of natural frequencies a system has depends upon the complexity of the system. A simple oscillator has only one natural frequency because the oscillator can vibrate in only one way. It follows that the number of natural frequencies a complicated system has depends on the number of different ways the oscillator can vibrate. Each of these different ways of vibrating is called a *normal mode,* and each normal mode has its own natural frequency. In addition to having a particular natural frequency, each normal mode also has its own damping time.

   When discussing the vibration of complicated systems, we must include yet another consideration: The system does not always vibrate in just one of its normal modes. However, as we will see, no matter how the system vibrates, its motion can be built up from a mixture of its normal mode motions. This mixture of normal modes in a complicated vibration is called the *vibration recipe* (just as the mixture of ingredients to make a cake is called its recipe).

Normal Modes

We will start our discussion of complicated vibrating systems by looking first at a simple vibrating system that is different from the ones discussed in Chapter 2. This simple vibrating system is made by fixing a bead (or weight) on an elastic string and tying the ends of the string to supports. We will consider the movement of the bead only in a direction perpendicular (or *transverse*) to the string. For this system, there is only one transverse mode of vibration, so the system is a simple oscillator. This motion is illustrated in Figure 3.1 where the solid line shows the equilibrium position, and the dashed lines show various positions of the string and weight as the system vibrates. When the bead moves away from its equilibrium position, the string stretches a little bit and, in stretching, pulls back on the bead. Because this system vibrates in only one way, it has only one natural frequency. The value of the natural frequency depends upon the elasticity of the string, how tightly the string is stretched, and the mass of the bead mounted on the string. This vibrating system also has a decay time, and we could measure the system's frequency and decay time in the same way that we measured them for the pendulum in the last chapter.

We consider a single bead on an elastic string because we then can move to successively more complicated systems by adding beads to the string. The first step in this process is to consider two beads on an elastic string. Having two beads instead of one provides greater variety in the way this system can move. Physicists describe this greater variety by saying each bead gives the system a *degree of freedom*. Thus, with one bead, we have one degree of freedom. In this example, the degree of freedom means the bead is free to move back and forth in a direction perpendicular to the string. The two-beaded string has two degrees of freedom because not only is one bead free to move back and forth perpendicular to the string (just like in the single-beaded system), but the second bead can move freely back and forth, independent of the first bead. For example, if the first bead is moving up, then the second bead could be moving up, moving down, or standing still. On the other hand, the motions of the beads do affect one another. For example, if one bead is sitting at its equilibrium position and the other bead has moved up, then this displacement of the second bead will cause the elastic string connecting the two beads to pull the first one up also. In spite of this interaction, however, a knowledge of one bead's motion does not permit us to determine the motion of the other bead; thus, there are two degrees of freedom.

In general, the motion of a two-beaded string appears very chaotic. Under certain circumstances, however, the string's oscillation will appear to be quite simple. In the first of two possible simple cases, the two beads move back and forth in phase

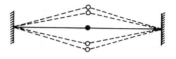

Figure 3.1. A vibrating bead on an elastic string

Weights on elastic strings. The upper single weight is a Styrofoam plastic ball, and the two weights on the lower string are lead fishing sinkers.

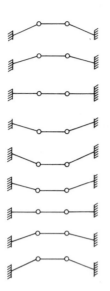

Figure 3.2. The first normal mode vibration of a two-beaded string

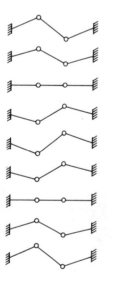

Figure 3.3. The second normal mode vibration of a two-beaded string

with each other. In the other simple case, the two beads move back and forth in antiphase with each other. These two cases are the *normal modes* of vibration for a two-beaded string and are illustrated with sequences of pictures in Figures 3.2 and 3.3. It is important to recognize a couple of facts about normal modes of vibration. First, when a system is vibrating in one of its normal modes, every part of the system is vibrating at the same frequency. Second, each normal mode has a different frequency of vibration. Thus, the time required for the two-beaded string to complete the vibration shown in Figure 3.3 is different than the time to complete the vibration shown in Figure 3.2. The time in each case is the period of the vibration, and different periods mean different frequencies. Each normal mode also has its own damping time. Finally, we must recognize that a system can vibrate in a way that is not one of its normal modes. However, as we shall see, any motion of the system can be described in terms of its normal mode motions.

In this book, I shall always label the normal modes by numbers, and the first normal mode will always have the lowest frequency. Thus, for the two-beaded string (with two degrees of freedom), there are two normal modes of vibration, and these two normal modes have frequencies of f_1 and f_2. The actual values of these frequencies depend upon the mass and location of the beads and the elasticity and tightness (tension) of the string. However, when identical beads are spaced equally along the string, the ratio of frequencies f_2/f_1 is equal to 1.74. Another way of saying this is that the frequency of the second normal mode as shown in Figure 3.3 is 1.74 times the frequency of the first normal mode, which is shown in Figure 3.2.

Now, let's consider what happens when more beads are added to the string. For a system consisting of an elastic string with three beads on it, there are three normal modes of vibration. For equally spaced beads, the normal mode patterns are shown in Figure 3.4 along with the ratios of the frequencies for each normal mode of vibration. Notice that in the second normal mode of vibration, the center bead does not move at all. A point like this that is not moving is called a *node*. (The second normal mode of vibration for a two-beaded string also has a node at the center of the string, although there is no bead located at that point.) The third normal mode of vibration for a three-beaded string has two nodes in it, but just as for the second normal mode of the two-beaded string, there are no beads at these nodes.

The four normal modes for a four-beaded string are shown in Figure 3.5. We can summarize the behavior of beaded elastic

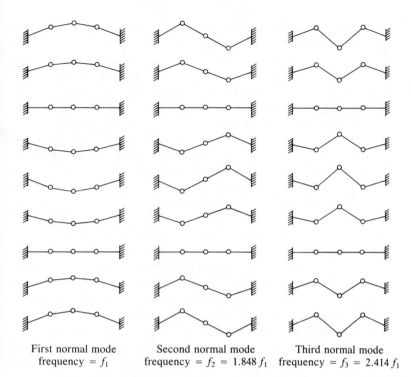

First normal mode
frequency = f_1

Second normal mode
frequency = f_2 = 1.848 f_1

Third normal mode
frequency = f_3 = 2.414 f_1

Figure 3.4. Vibration patterns of a three-beaded string (Diagrams show equal phase increments, not equal time increments.)

strings by noting the following similarities among the behavior patterns for them:

1. The number of different modes of vibration is equal to the number of beads on the string (which in turn is equal to the number of degrees of freedom of the system).

2. When a system is vibrating in its first normal mode, all parts of the system are in phase with one another.

3. For the normal mode numbered *N*, there are always *N–1* nodes (not counting the ends).

4. As the number of beads on the string increases, the frequency of the second normal mode of the system becomes closer and closer to twice the frequency of the first normal mode.

Figure 3.5. The vibration patterns of a four-beaded string

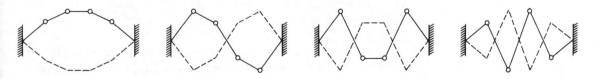

Now, we want to consider the normal modes of a string without any beads on it at all. We will call such a system a *continuous string* or just a *string*. One useful way of looking at this system (because of our work with beaded strings) is to consider the string as having a large number of very small beads on it. When I say a large number of beads, I mean thousands of them. If a string had thousands of beads on it, it would have thousands of degrees of freedom and, therefore, thousands of normal modes of vibration. We will not be interested in looking at all of these normal modes, only the first few. Just as with the beaded strings we have looked at, the continuous string will have all its parts vibrating in phase with each other when it is vibrating in its first normal mode. When the string is vibrating in its second normal mode, there will be one node at the center of the string. When the string is vibrating in its third normal mode, two nodes will be between the ends, and so forth. When a continuous string is perfectly flexible, we will call it an *ideal continuous string*. For such a string, a very special relation exists among the frequencies of the normal modes. The frequency of the second normal mode is exactly twice the frequency of the first normal mode, and the frequency of the third normal mode is exactly three times the frequency of the first normal mode. In general, the frequency of the normal mode numbered N is N times the frequency of the first normal mode. Since N is an integer, we can describe this special situation by saying the frequencies of the higher modes of vibration are integral multiples of the first normal mode frequency. We can write this concisely as

$$f_N = Nf_1 \qquad (3.1)$$

We could have anticipated this result by extrapolating the fourth point in the above summary for beaded elastic strings to the present case of a large number of small masses on the string. When this special situation (where the frequencies of the higher normal modes are integral multiples of the frequency of the first normal mode) occurs, we call the normal modes by a new name: *harmonics*. Thus, for an ideal continuous string, we can talk in terms of either normal modes or harmonics. However, for a beaded string or a non-ideal continuous string (which we will discuss later), we can talk only in terms of normal modes.

In addition to the terminology of normal modes and harmonics, there is a third way to describe how complicated systems vibrate. The normal mode with the lowest frequency is termed the *fundamental*. All the other normal modes of vibration are called *overtones*. Just as we label the normal modes with a number, we label the overtones with a number. Since the first normal mode is called the fundamental (and not an overtone), the first overtone is the second normal mode. This correspondence continues for all the other normal modes of the system. That is, the third normal mode is the second overtone, etc. This way of describing the modes of vibration can be applied to all vibrating systems. Only the harmonic description is limited to certain types of systems.

Some new, interesting aspects exist regarding the shape of a completely flexible or ideal continuous string that is vibrating in one of its normal modes. The shapes of such a string for the first nine harmonics are shown in Figure 3.6. In each case, the shape of the string is a sine curve. The drawings of the string in Figure 3.6 look just like the sine curves we observed on the oscilloscope screen for simple vibrations. While the two sets of curves look the same, we must recognize that they show two different things. The curve on the oscilloscope screen shows how the position of a single object (or a single point on a string) varies with time. The pictures in Figure 3.6 show what the string looks like at a single instant of time. In other words, the picture of the curved string shows how the displacement of the string varies with the distance along the string. Thus, the first sine curve (that on the oscilloscope screen) shows how displacement at a single point varies with time, while the second sine curve (that of the curved string) shows how displacement varies with position along the string at a single instant of time. To use a photographic metaphor, the first curve is a movie of a single point, and the second curve is a snapshot of the entire string.

Whenever the string vibrates in one of its normal modes, each point on the string executes simple harmonic motion. The amplitude of this motion varies with position along the string. The amplitude at some points is zero. Just as with the beaded string, we call these points *nodes,* and the number of nodes between the ends of the string depends on the normal mode of vibration. That is, a string vibrating in the first normal mode will not have any nodes between its ends; a string vibrating in the second node will have one node between its ends, and so forth. At other points, the amplitude of vibration is a maximum. These points are called *antinodes.* The number of antinodes appearing on the string also depends upon the normal mode in which the string is vibrating. In Figure 3.6 it's easy to see that for any normal mode of vibration, the number of antinodes that occurs is always one more than the number of nodes existing between the ends of the string. Thus, we have a very easily

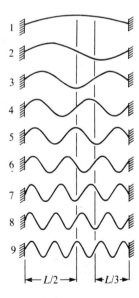

Figure 3.6. The first nine normal modes of vibration for a perfectly flexible string

remembered relationship: The number of antinodes appearing on the string is equal to the number of the normal mode (or harmonic). For example, the third harmonic of a vibrating flexible string always has three antinodes and, therefore, two nodes between ends of the string.

Superposition of Normal Modes

When a system is vibrating in one of its normal modes, the appearance of the system repeats itself in a very simple, regular way. But, a system does not always vibrate in one of its normal modes. An example of such behavior is shown in Figure 3.7 with the relatively simple two-beaded string. In this example, the right-hand bead initially was pulled upward and then released; subsequent positions of the two beads at equal time intervals are shown. As can be seen, the motion is very complicated. We would have to wait a very long time to see the motion repeat itself, and for real systems, the motion would die away before this happened. There is, however, a simple way of visualizing why the beads move the way they do. This way of looking at the system involves the superposition of normal mode motions.

Before explaining the motion shown in Figure 3.7 using normal modes, we first must recall that superposition refers to the idea that the effect we see is the sum of two separate, individual effects. Using this idea, we will describe the motion of the two-beaded string shown in Figure 3.7 in terms of the string's normal modes. In this case, superposition tells us that the displacement of each bead at each instant of time will be the sum of the displacements that would be produced by each normal mode separately. The size of the displacement produced by each normal mode depends upon the amplitude of the normal mode vibration and where the normal mode vibration is in its cycle (that is, its phase). The amplitudes and phases of the two normal mode vibrations for the two-beaded string in Figure 3.7 are determined by the way the system's vibration is started. Physicists term this information *initial conditions*. For the case shown in Figure 3.7, the right bead was pulled up a distance, D, and then released. When we pull the right bead up the distance, D, the left bead follows along and rises a distance $\frac{1}{2} D$. Thus, we have the initial conditions: The right bead starts from rest at a distance, D, above the equilibrium position, and the left bead starts from rest at a distance $\frac{1}{2} D$ above the equilibrium position. We now will use this information to determine the amplitude of vibration for each normal mode. For this particular case, the amplitude for

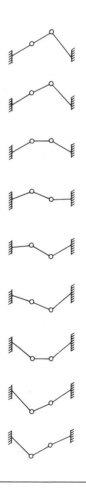

Figure 3.7. The movement of a two-beaded string that is not vibrating in one of its normal modes

the first normal mode is ¾ D, and the amplitude for the second normal mode is ¼ D. Since both beads start from rest, we also know that the two normal modes are in antiphase when the motion starts. However, the relative phase between the normal modes will change continuously because the two modes have different frequencies. Figure 3.8 shows how the displacements of the two normal modes add together to give the displacement of the string.

Now, let's assume we have an ideal system, which means no damping exists. In this case, when the right bead is released and the system starts vibrating, the amplitude of each normal mode's vibration will remain constant. (Note that this is a beaded string, so when we speak of an ideal system, we do not mean to imply that harmonics will be present. Harmonics are present only with ideal continuous strings.) At every instant of

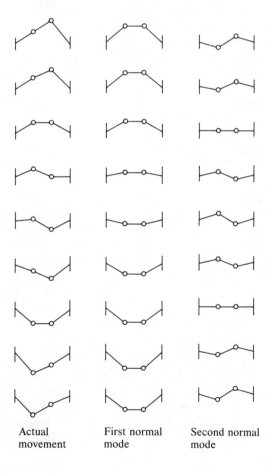

Actual
movement

First normal
mode

Second normal
mode

Figure 3.8. The movement of the two-beaded string in Figure 3.7 shown as a superposition of normal mode motions

time, the displacement of each bead will be the sum of the two displacements that would exist for each normal mode. The resulting vibration appears so complicated because the frequencies of the two normal modes differ. The vibration of the two-beaded system shown in Figure 3.7 is repeated in Figure 3.8 along with the vibrations of each normal mode. We need to emphasize several points about this construction, including

1. The higher frequency of the second normal mode means the second normal mode moves through its cycle more rapidly than the first normal mode.

2. At each instant of time, the actual displacement of each bead is the sum of the displacements the bead would have according to each normal mode by itself.

3. At certain instants of time, the two normal mode displacements cancel (or partially cancel) each other because one is trying to move a bead up, and the other is trying to move the same bead down. For example, look at the left bead in the very first picture. The first normal mode is pulling the bead up, and the second normal mode is pulling the bead down.

Superposition also can be used to describe the shape of a continuous string in terms of the string's normal modes. For this situation, however, there are two important differences. First, there are many modes of vibration, so the shape of the string at any instant of time can be very complicated. In contrast, the second difference simplifies the behavior. As you will recall, we have harmonics for an ideal string. Therefore, the frequencies of all the normal modes are integral multiples of the fundamental frequency, which means that during the period of the fundamental mode of vibration, every other mode has completed an integral number of cycles. Thus, the shape of the string repeats itself during every cycle of the fundamental mode.

An example of an ideal string vibrating with a mixture of normal modes is shown in Figure 3.9. In this figure, the shape of the string is followed through one-fourth of its cycle. During the rest of the cycle, which is not pictured, the string and the normal modes will repeat the shapes shown here (although for the next half of the period, the shapes will be turned upside down). At each step, shapes of the normal modes are being added to give the actual displacement of the string. For the particular case shown here, we have used only three normal modes to produce the string's displacement. We also have chosen the relative

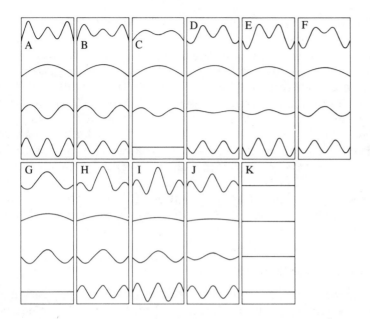

Figure 3.9. A flexible string vibrating in a combination of the first, third, and fifth harmonics. Pictures A–K show the shapes of the string and each harmonic at equally spaced times. The top line in each picture is the shape of the string, and the other three lines are the shapes of each harmonic.

amplitudes and phases of the three vibrations. The amplitudes of the third and fifth harmonics are sixty percent and eighty percent the size of the fundamental mode's amplitude, respectively. Also, the three vibrating modes are in phase with each other in picture A. Just as in the case of the vibrating two-beaded string described earlier, the higher modes of vibration oscillate more rapidly than the fundamental mode does. Thus, during the period of time covered by the pictures in Figure 3.9, the fundamental mode has completed one-fourth of a cycle. The third harmonic has completed three-fourths of a cycle, however, because it is oscillating three times faster. And the fifth harmonic is oscillating at five times the frequency of the fundamental mode, so the fifth harmonic has completed one and one-fourth cycles.

The particular choice of normal mode amplitudes and phases used to construct the vibrating string of Figure 3.9 illustrates an interesting aspect of a general vibrating string. There are no nodes for the actual string. This fact is evident because no points on the string remain stationary (except the ends, of course, and I am not considering them in my counting of nodes).

Nodes easily could exist for other choices of normal modes. For example, suppose a complicated vibration were built up using only even harmonics (the second, fourth, sixth, etc.). Each of these harmonics has a node at the center of the string. (See

Figure 3.6 for some examples.) Thus, when many even harmonics are superposed (that is, added together), none of them will cause any displacement at the string's center. Consequently, a node will be located at the center of such a vibrating string.

A similar thing happens when a vibration is built only from harmonics whose number is divisible by three. Each of these harmonics has a node located at points one-third of the way along the string from each end. (Again, this finding may be checked for the first few cases by looking at Figure 3.6.) Of course, the higher harmonics such as the sixth, ninth, etc., have more nodes, but these harmonics do have the two nodes mentioned above. Thus, the superposition of these harmonics (with any amplitudes and phases) will have nodes located at those points.

Vibration Recipe

As mentioned, any complicated vibration is the superposition of normal mode vibrations. The actual vibration depends on the relative amplitudes and phases of the various normal modes comprising the vibration. A convenient and concise way of describing the relative amplitudes of each normal mode in a vibration is to use a bar graph as shown in Figure 3.10. In this graph, the normal mode number is plotted along the horizontal axis, and the amplitude is recorded in the vertical direction. Thus, the height of each line shows the amplitude of each normal mode's vibration. It is also possible to plot the frequency along the horizontal axis. Since the string in Figure 3.9 is ideal, its normal modes are harmonics, so the three equally spaced vertical lines in Figure 3.10 also are spaced equally in frequency. Whenever the normal modes are harmonics (as with an ideal string), there is no difference between the two plots. However, in the cases of beaded strings or non-ideal continuous strings, there is a difference. Plots done using normal mode numbers still will have lines equally spaced, while plots using frequencies will not.

Notice that the graph in Figure 3.10 contains no information about the relative phases of the three normal mode vibrations. We shall see in Chapter 5 that such relative phase information is not important for sound vibrations because the ear cannot detect any difference in phases. Therefore, the graph in Figure 3.10 gives as much information about complicated sound vibrations as we can use. This type of graph will be very

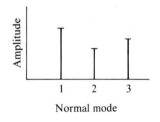

Figure 3.10. Vibration recipe for the vibrating string shown in Figure 3.9

helpful in Chapters 4, 8, and 9 when we describe the differences between various musical sounds.

In the examples considered above, we were looking at ideal systems that had no damping. For a real vibrating system, each normal mode would have damping that is characterized by a damping time. In general, the damping times of the normal modes will be different. For vibrating strings, either beaded or continuous, the higher modes of vibration usually have shorter damping times. Thus, if we start a string vibrating in such a way that many normal modes are involved, the string eventually will end up vibrating in its fundamental mode only.

Starting a String Vibrating

There are basically three ways to start a string vibrating in musical instruments: hitting, bowing, or plucking, and we will take a brief look at each method. The first method, hitting, is used in a piano where a hammer is forced against a string when the piano key is depressed. Of course, we want the hammer to move quickly out of the way so it will not interfere with the string's rebounding after being hit. The string moves back in a time interval that is one-half the vibration period, and for the higher-pitched notes, this is a very short time, so the hammer's motion must be very quick! Each instrument in the violin family is played using primarily the second method to move a string, called bowing. The part of the bow that is drawn across the string is made of horsehair, which has a very rough surface. As the bow moves, the horsehair grabs the string, pulls the string a short distance, releases it, and grabs it again. The last method, plucking, occasionally is used with the violin, but this technique is more strongly associated with guitars (both acoustic and electric), banjos, mandolins, ukeleles, and the harpsichord. When a string is plucked, it is pulled strongly to one side and then released.

In each of these ways of driving a string, different vibration recipes will be produced, causing different tonal qualities of the notes being played. But the resulting vibration recipe also depends on the point along the string where the hitting, bowing, or plucking takes place. For simplicity, I will use the term *driving* to refer to the process of starting the string's vibration. Thus, driving can be hitting, bowing, or plucking. An extreme example serves as an excellent illustration. Driving the string at its center means that there will be at least some displacement at that point. Thus, any normal mode that has a node there cannot be included in the resulting vibration recipe. As we saw earlier, all the even harmonics have a node at the string's center. Therefore, a string driven at its center will have only odd harmonics being superposed to create the vibrating string. There is no way to predict the relative amplitudes of these odd harmonics, and different vibration recipes are produced by hitting, bowing, and plucking. But we do know that none of the methods will excite any of the even harmonics.

A similar type of behavior occurs when the string is driven at a point one-third of the way along the string. Again, as we saw earlier, every third harmonic has a node at this point. Thus, in the vibration recipe for a string driven at this point, there will be no contribution from these harmonics (the third, sixth, ninth, etc.). Figure 3.11 shows possible vibration recipes for strings driven at their center and at a point one-third of the way in from one end.

In the lab, there are other ways to start a string vibrating. One of these, first used by Franz E. Melde (in 1859) and, therefore, called Melde's experiment, has the string attached to a tuning fork. The string, driven by the tuning fork, vibrates at the fork's natural frequency. By suitably adjusting the length and tightness (or tension) of the string, we can make the string vibrate in one of its normal modes. Two different experiments can be performed with this setup. In the first, we keep the string tension constant and vary only the length of the string (remem-

Figure 3.11. Vibration recipes for strings driven at different points

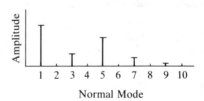

A. Driving point at center

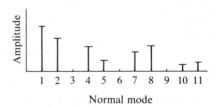

B. Driving point one-third from one end

ber the frequency is always that of the tuning fork). We find that for certain lengths, we get a normal mode vibration with a large amplitude. The length is held constant, and the string tension is varied in the second experiment. An easy way of doing this is to hang the end of the string over a pulley and suspend some weights from it. We again find large amplitude normal mode vibrations for certain values of the tension. These two experiments are pictured in Figures 3.12 and 3.13. There are several things to note in these pictures.

1. The tuning fork is not located exactly at a node of the string's vibration. The reason, of course, is that the vibrating tuning fork (and attached string) has a small, but non-zero, amplitude.

2. When the tension in the string does not change (as in Figure 3.12, the distance between adjacent nodes on the string is the same. (We will learn why this situation happens in Chapter 8.)

3. When the length of the string remains fixed, a smaller tension produces a higher mode of vibration. (This condition also will be explained in Chapter 8.)

When the amplitude is large in these experiments, the string is resonating; that is, there is a large transfer of energy from the driver (the tuning fork) to the vibrating system (the string). In Chapter 2, we discussed resonance from a slightly different point of view. In those cases, we had vibrating systems that were not changing their characteristics. Thus, the natural frequencies of the vibrating system were fixed, and resonance occurred when the varying frequency of the driver matched the system's natural frequency. Here, in contrast, the driving frequency is fixed, so the natural frequency of the system has to be changed until it matches that of the driver.

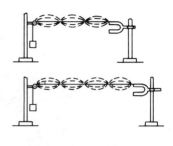

Figure 3.12. Melde's experiment with both tension and frequency held fixed (The length is varied.)

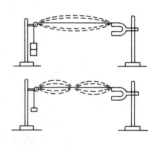

Figure 3.13. Melde's experiment with both length and frequency held fixed (The tension is varied.)

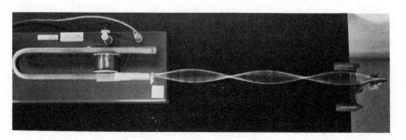

Melde's experiment. The electrically driven tuning fork is vibrating the string at a frequency of 80 Hz, and the string is vibrating in its third normal mode.

We also can use a variable frequency device to drive a string in the lab. In this experiment, we keep both the string length and the string tension fixed and vary the driving frequency. Thus, the characteristics of the vibrating system stay the same. Now we find the same type of resonance behavior we saw in Chapter 2. That is, as the natural frequency of each normal mode is matched by the driver's frequency, the amplitude of the string's vibration increases dramatically. For a nice flexible string, these frequencies will be integral multiples of the fundamental frequency. With this type of experiment, we can develop a response curve just as we did for simple systems in Chapter 2. Figure 3.14 shows the response curve for a typical ideal string. Each normal mode of the string has a peak on the response curve. The height of the curve shows the resulting vibrational amplitude of the string when driven at that frequency. Weaker resonances occur when the peaks are smaller. The widths of the peaks also vary. As noted in Chapter 2, wider peaks occur for stronger damping. Thus the normal modes with narrow peaks have longer damping times. For the response curve in Figure 3.14, we see that the first normal mode has the strongest resonance and the longest damping time. Each successively higher mode has a weaker response and a shorter damping time. Also, since the normal modes are harmonics for this ideal string, the peaks are spaced equally along the horizontal axis.

Thus, there are several ways to study vibrating strings. When we use a fixed frequency driver, we must change the characteristics (tension or length) of the string to have a normal mode vibration. If, on the other hand, we keep the string (and its length and tension) the same, we must change the driving frequency to excite the normal modes. The reasons behind these observations will be explained in Chapter 8.

Figure 3.14. The response curve for a typical ideal string

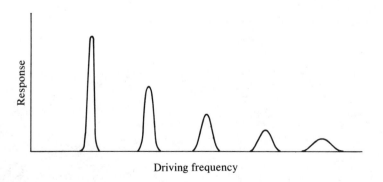

Vibrating Surfaces

Thus far, we have limited our discussion of complicated vibrating systems to beaded strings and ideal continuous strings. We shall use the information we have gained about the latter when we discuss stringed instruments in Chapters 4, 8, and 9. There are other complicated vibrating systems we must mention, however. The first is a column of air inside a pipe. This vibrating system is very similar to the vibrating string and will be discussed later in connection with wind instruments.

An even more complicated vibrating system is a surface such as a drum head or flat plate. The concepts we developed for describing the vibration of a string can be carried over to the case of a vibrating surface, but there are a few complications. First, a surface can vibrate in many more ways than a string can, so there are many more normal modes of vibration. In addition, the normal modes are not harmonic, so the frequencies will not be some integer multiple of a fundamental frequency. A third interesting complication is that nodes are no longer points as they were on a string. Rather, they are lines on the surface. Thus, we speak of *nodal lines;* along a nodal line, there is no motion of the surface. The antinodal points of a vibrating string also may become lines when the vibrating object is a surface.

The nodal lines can be observed by performing an experiment first done by Ernst Friedrick Chladni in the latter part of the eighteenth century. In this experiment, a thin flat metal plate is clamped at its center, and a violin bow is used to make the plate vibrate. The bow is drawn across the edge of the plate, causing the plate to vibrate in the same way that the violin string does when bowed. Sand sprinkled on the plate bounces away from the plate's vibrating parts and collects along the nodal lines. We can force the plate to vibrate in different normal modes either by changing the point at which it is bowed or by touching the plate at different points while it is bowed. Wherever we touch the plate, there will be a node because we are preventing the plate from moving. From such a point, a nodal line will extend across the plate. In the same way, wherever the plate is bowed, there will be an antinode and the start of an antinodal line. We cannot see these antinodal lines because the sand bounces away from them. However, there always will be an antinodal line between each pair of nodal lines.

There is another interesting observation we can make during a Chladni plate experiment. The vibrating plate produces a sound, and, just as with any sound, its pitch corresponds to the

Chladni plate patterns. The method of using a bow to vibrate the plate to produce the patterns is shown in A. Different patterns are shown in B through D. In all patterns, the sand has collected along the nodal lines. The lines across each figure represent the edge of the table on which the Chladni plate clamp is mounted.

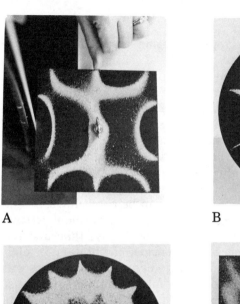

A

B

C

D

vibration's frequency. Now, the observation we can make is that when we excite different normal modes of the plate, we get different sounds. The closer together the nodal lines are on the plate (and therefore the higher the normal mode of vibration), the higher the pitch of sound produced. We would expect this relationship from our discussion of the behavior of a vibrating string.

The Production of Sound by a Loudspeaker

We have seen how vibrating objects serve as sources of sound. The way the object vibrates determines the characteristics of the sound. The vibration of a tuning fork, for example, is simple harmonic motion, so the sound produced is a pure tone. Vibrating strings and (as we will see in Chapters 8 and 9) columns of air in wind instruments vibrate in more complicated ways through a superposition of their normal mode vibrations. (Each normal mode vibration is simple harmonic motion of a different frequency.) In any of these cases, vibrations are transferred to the air, pass through it, and are transferred to our eardrum. In Chapter 6, we will see how the ear detects these vibrations so they can be interpreted by the person hearing them. Instead of passing the vibrations to an ear for detection, the vibrations could be sent to and detected equally well by a microphone. The microphone changes the mechanical vibrations into a vibrating electrical signal. In Chapter 1, we saw how this signal can be displayed on the screen of an oscilloscope, but the signal could be stored as easily on a magnetic tape or a phonograph record. Following the recording of these sounds, they can be rebroadcast using a loudspeaker. To rebroadcast these sounds, the vibrating electrical signal is reconstructed from the tape recording or phonograph record and is used to drive a loudspeaker. Thus, the loudspeaker produces vibrations in the air that are identical to the ones originally detected and recorded by the microphone.

Since the loudspeaker is another source of sound, we will investigate how it works. A cross-sectional drawing of one is shown in Figure 3.15. The *frame* of the loudspeaker is rigid and, in practice, is attached to an enclosure or cabinet. At the rear of the speaker is the *driver,* a device that converts electrical signals into mechanical vibrations. The part that actually serves as the sound source in the loudspeaker is the *cone.* At its large end, the cone is attached—by means of flexible material—to the frame. In order to reproduce the sound that originally was recorded, the cone of the speaker is moved forward and backward by the driver, recreating the original vibrations. One difficulty in this task results from the wide range of frequencies that are present

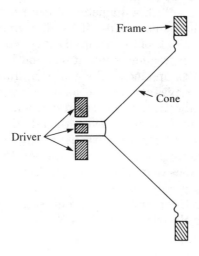

Figure 3.15. A cross-sectional view of a loudspeaker

Figure 3.16. Vibration recipes

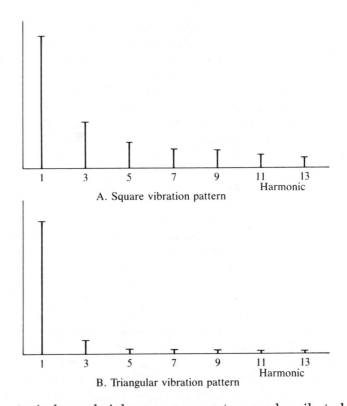

A. Square vibration pattern

B. Triangular vibration pattern

in a typical sound. A large cone cannot respond easily to high frequency vibrations, and a small cone does not couple effectively to the air for low frequency vibrations. Audio engineers have solved this problem by designing speaker systems that include several different sizes of loudspeakers. A *woofer* has a diameter of about 25 cm and is very effective in reproducing low frequency sounds. With a diameter of about 8 cm, a *mid-range* speaker earns its name because it handles vibrations with frequencies in the middle of the required range. Finally, a *tweeter* is designed for high frequency vibrations and has a diameter of about 3 cm. When an electrical signal containing the information of a sound arrives at such a speaker system, there is a device that sorts the vibrations of various frequencies and sends them to the appropriate speaker. Even though the vibrations are sorted into different ranges of frequencies, those reaching each speaker are usually not sine curves. Rather, the vibrations are complicated shapes that reflect the structure of the sounds produced by the original sources.

As noted in Chapter 2, an electronic signal generator can be used to produce electrical vibrations. When these vibrating electrical signals are fed to a speaker, the appropriate movement is passed to the speaker's cone. In this way, the pure tones associated with sine curve vibrations can be generated. Also, the other patterns shown in Chapter 2, such as the square, sawtooth, and triangular vibration patterns, can be broadcast as sound. Each of these patterns has a different tone quality. One way of understanding this difference is to recognize that each of these patterns has a different vibration recipe. The vibration recipes for the square and triangular vibration patterns are shown in Figure 3.16. In the same way, a musical instrument or the human voice has a unique vibration pattern and a corresponding vibration recipe. For any of these vibration patterns, a loudspeaker (or, more correctly, a loudspeaker system) can perform the correct vibrations and thereby can create the desired sound.

Summary

In this chapter, we have looked at several vibrating systems that we classify as complicated because they have more than one normal mode of vibration. For this reason, any extended object is usually a complicated vibrating system. In general, such a system has the same number of normal modes of vibration as it has degrees of freedom. A beaded elastic string (which is a good complicated system to start with) has the same number of normal modes and degrees of freedom as the number of beads. In this regard, a string with no beads can be considered as an elastic string with thousands of very small beads that are very close together.

We can refer to the various normal modes of a complicated vibrating system in several different ways. In the simplest, we number these normal modes in ascending order according to their frequencies. Another way is to name the first normal mode as the fundamental; the higher normal modes are overtones, again numbered in ascending order. Finally, when the higher mode frequencies are integral multiples of the fundamental frequency, we can call the normal modes harmonics, and we distinguish between the various harmonics by numbering in the same way we did for the normal modes.

When a complicated object vibrates in one of its normal modes, every part of the object executes simple harmonic motion with the same frequency. The amplitudes of these motions are not the same, however. Those places that do not move are called nodes, and those having large amplitudes are called antinodes. For a vibrating string, the number of the normal mode corresponds to the number of the antinodes in the vibration. For a vibrating surface, such as Chladni plate, the nodes and antinodes are lines rather than points.

Extended objects also can move in more complicated ways, and these motions can be explained as superpositions of normal mode motions. The vibration recipe describes the amount of each normal mode in such a complicated vibration. When a string is driven (by bowing, plucking, or hitting) at a particular location, the recipe of the resulting vibration will not contain any normal mode that has a node located at the driving position.

Suggested Readings

The material in this chapter also is covered by both books suggested in Chapter 2: *Horns, Strings, and Harmony,* by Arthur H. Benade (Garden City, N.Y.: Anchor Books/Doubleday, 1960) and *Exploring Sound* by Alexander Efron (New York: Hayden Book Company, 1969). In particular, beads on elastic strings are discussed extensively by Benade.

An interesting example of a complicated vibrating system is discussed in the article "Resonant Vibrations of the Earth" (*Scientific American.* November 1965, pp. 28–37). The article describes different modes of vibration for our planet and how the measurement of the vibration recipe following an earthquake is helping geologists learn more about the structure of the earth.

Review Questions

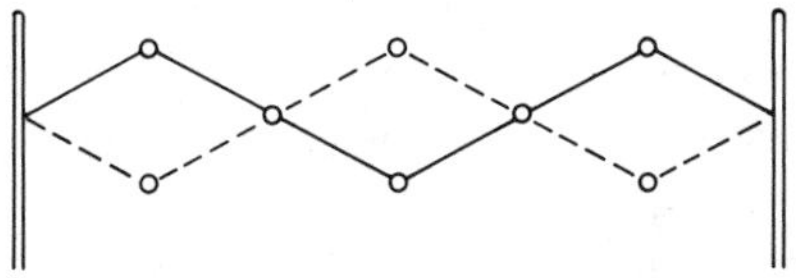

Figure 3.17. For use with review question 1

1. In which normal mode is the five-beaded string in the Figure 3.17 vibrating?

    a. First

    b. Second

    c. Third

d. Fourth

e. Fifth

2. Five beads are equally spaced along an elastic string. If we measured the frequencies of all the normal modes, we would find that the frequency of the first overtone _____ the frequency of the fundamental.

a. is equal to

b. is between one and one-half and one and three-fourth times

c. is between one and three-fourth and two times

d. is exactly two times

e. has no predictable relation to

3. How many antinodes does a string (an ideal continuous one) that is vibrating in its fifth overtone have?

a. One

b. Four

c. Five

d. Six

e. Twelve

4. We drive a string at 450 Hz and see five antinodes. Therefore, the frequency of the fundamental is

a. 50 Hz.

b. 90 Hz.

c. 150 Hz.

d. 450 Hz.

e. 2250 Hz.

5. The frequency of a string's fundamental is 100 Hz. What is the frequency of the third overtone?

a. 25 Hz

b. 33 Hz

c. 50 Hz

d. 300 Hz

e.   400 Hz

6.   A 350 Hz tuning fork is used in a Melde's experiment with a string having a fundamental frequency of 70 Hz. In which overtone and harmonic will the string be vibrating?

a.   Fifth and fourth, respectively

b.   Fourth and fifth, respectively

c.   Fifth and sixth, respectively

d.   Sixth and fifth, respectively

e.   Fifth for each

Figure 3.18.   For use with review question 7

7.   A three-beaded string is vibrating in a superposition of its normal modes. At a particular instant of time, the three nodes have the displacement shown in Figure 3.18. At this same instant, the string actually has the shape shown by which of the following?

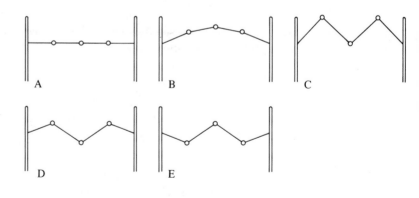

a.   A

b.   B

c.   C

d.   D

e. E

8. On a violin, the length of the string is 40 cm, and the bow is drawn across the string 5 cm from one end. Therefore, we expect the vibration recipe of the string will not have the _____ overtone in it.

    a. fourth

    b. fifth

    c. sixth

    d. seventh

    e. eighth

9. A 60 cm string driven by a tuning fork is vibrating in its third overtone. How long must the string be (keeping the tension the same) in order for it to vibrate in its fifth overtone?

    a. 75 cm

    b. 90 cm

    c. 100 cm

    d. 120 cm

    e. 180 cm

10. The vibration recipe of a plucked string looks like Figure 3.19.

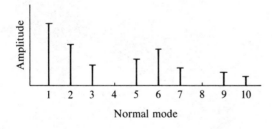

Figure 3.19. For use with review question 10

The string was probably plucked _____ of its length from one end.

    a. one-half

    b. one-third

c. one-fourth

d. one-fifth

e. one-twelfth

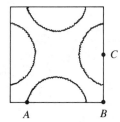

Figure 3.20. For use with review question 11

11. The pattern of sand on a Chladni plate is shown in Figure 3.20. This pattern could result from

   a. bowing the plate at A and holding it at B.

   b. bowing the plate at C and holding it at B.

   c. bowing the plate at A and holding it at C.

   d. bowing the plate at C and holding it at A.

   e. bowing the plate at B and holding it at C.

4

The Piano as a Source of Sound

Important Concepts

- A piano is a fixed-frequency instrument that is capable of producing eighty-eight different tones.

- The basic vibrating object in the piano is a string; its vibrations are transferred to the air by means of a sounding board, which is coupled to the string by the bridge.

- The fundamental frequency of each string is determined by its length, tension, and weight.

- Since piano strings are stiff, their normal modes are not harmonics; higher normal modes have frequencies slightly larger than corresponding harmonics.

- The vibration recipe of a struck string varies with time. Usually, higher normal modes have shorter damping times.

- The fundamental interval in any musical scale is the octave for which the frequency doubles.

- Two tones separated by a just interval have frequencies that are proportional to a ratio of small integers.

- A natural diatonic scale contains only just intervals.

A grand piano (Picture provided courtesy of the Kimball Keyboard Division.)

- The frequency ratio for every pair of adjacent tones in an equally tempered scale is about 1.06.

As we have seen, vibrations produce sounds. Some vibrations produce sounds that we call music, and it is interesting to see what is special about these sounds and vibrations. One aspect of our discussion will be to look at how a number of different musical instruments operate. The piano is an ideal instrument for starting this discussion for several reasons: the same sound is repeated easily on a piano; the musical notes produced by the piano correspond to the musical scale used in western culture; and the piano is a complicated instrument, so many aspects of musical sound production are illustrated with it. In addition, the basic vibrating system of the piano is a string, which nicely follows our discussion in the last chapter.

Although the piano is not a standard instrument in an orchestra, the piano is used widely by itself or in combination with other instruments for musical performances. Eighty-eight different tones can be played on a piano by depressing a key and causing a hammer to strike a string. The frequencies of each string cannot be changed while playing the piano, so we call the piano a *fixed-frequency* instrument. When the piano is tuned, however, each string's frequency is adjusted so that it produces the desired tone that corresponds to a note of a musical scale.

The notes of the musical scale are named by using the first seven letters of the alphabet together with—in some cases—one of two special symbols, the sharp (♯) or flat (♭). These symbols are appended to a letter and define an increase or decrease in frequency, respectively. The entire range of musical frequencies is broken up into octaves; the frequency doubles for each rise in pitch of one *octave*. Within each octave, there are twelve notes $(A, B♭, B, C, C♯, D, E♭, E, F, F♯, G, G♯)$. Figure 4.1 shows the piano keyboard design for one octave. The white keys play the notes with unmodified letter names, while the notes with sharps or flats are played with the black keys. The fundamental frequency of the lowest note on the piano (an A) is 27.5 Hz, while that of the highest note (a C) is 4186.0 Hz.

Two notes near the middle of the piano's range deserve special mention. The first of these is *middle C*, which is the fortieth note from the bottom. This note is not in the exact middle of the piano's range; the note gets its name from the fact that in written music, it occurs midway between the treble and bass clefs. Thus, notes below *middle C* fall into the bass clef range, while

An upright piano (Picture provided courtesy of the Kimball Keyboard Division.)

notes above *middle C* are in the treble clef range. (See Figure 4.4). The second important note in the middle part of the piano's range is the *A* above *middle C*. This note has a frequency of 440 Hz and is used as the standard for pitch in western music.

In this chapter, we first will look at the physical construction of the piano. We will determine how each of the piano's components is used to produce musical sound. This study will include a discussion of how piano strings differ from the ideal strings considered in the last chapter. Then, we will look in detail at the actual sound produced by the piano for one of its notes. We conclude the chapter with a brief look at the physical basis for musical scales, and their historical development.

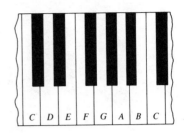

Figure 4.1. One octave of the piano keyboard

The Piano as a Vibrating System

As mentioned above, the basic vibrating system in a piano is a string. The frequency of the string's fundamental mode of vibration is what determines the pitch of the note. The fundamental frequency depends, in turn, on three properties of the string: its length, tension, and weight*. When any one of these three properties is changed (and the other two are held fixed), the frequency of all the normal modes of the string's vibration will change. If a string is made longer, then the frequency of all normal modes will decrease. If the weight (essentially the thickness) of the string is increased, then the frequency also will decrease. On the other hand, if we increase the tension of the string (that is, pull the string tighter), then the frequency also will increase. Thus, we see that two of these properties change the normal mode frequencies oppositely; whereas, a change in the third property affects the normal mode frequencies in the same way. That is, *increasing* either the length or weight of the string *decreases* the string's frequencies while *increasing* the tension *increases* its frequencies. All three of these properties are different for different strings in the piano, so different frequencies are produced. The strings that generate the lower notes of the piano are long and heavy and have a low tension. The strings producing sound in the intermediate part of the musical range of the piano have thinner, tighter strings that also may be shorter. Finally, those strings creating the higher notes are thinner, tighter, and shorter than the other strings in the piano.

---

*The technically correct term for the "weight" is *linear mass density* (or mass per unit length). However, strings that have high linear mass densities are usually thicker and heavier, so usually we can think of this property in terms of weight.

From a musical point of view, it is desirable for each piano string to have the same vibration recipe. That is, for every string, the bar graph showing the amplitude of each normal mode plotted against normal mode number should look the same. To achieve this, each string should be made of the same material and should have the same tension; only the length should change. But achieving this would require too great a variation in the length of the strings in order to cover the desired range of notes. For example, suppose the string for the highest note on the piano had a length of 5 cm (about two inches). Then, to have a string with the same thickness and tension produce the lowest note, the string would have to be 7.61 meters (about 25 feet) long. This length is obviously impractical, so compromises are made: Both the string tension and thickness are varied.

The way in which the weight (or thickness) is changed is particularly interesting. Piano strings actually are made of wire and, therefore, have some stiffness. If we obtain the added weight needed for low notes by increasing the wire diameter, then the bigger wires would get very stiff. As we will see shortly, this stiffness changes the frequencies of the overtones and is, therefore, an unacceptable alternative. Piano builders have solved this problem by wrapping a second wire (and, for the very lowest notes, a third wire) around the original one. This additional wire adds to the wire's weight, of course, but it does not make the wire too stiff.

All the piano strings are mounted on a strong rigid frame and pass over a bar that is called the *bridge*. One end of each string is attached firmly to the frame, and the other end of the string is wrapped around a peg that is attached to the frame and can be turned to vary the tension of the string. By changing the tension (and therefore the frequency), the string can be tuned to produce the proper pitch. A diagram showing the arrangement of a string, the bridge, and the sounding board (discussed in the next paragraph) is shown in Figure 4.2.

A string vibrating by itself in a room would not be heard very easily because the string's vibrations would not be transferred effectively to the air. A piano and all other stringed instruments improve their ability to transfer vibrations to the air by having a *sounding board*. The piano's sounding board is a flat piece of wood that is as big as the piano itself. For a violin, cello, or acoustic guitar, the body of the instrument is the sounding board. On all these instruments, the connection between the vibrating string and the sounding board is the bridge. As mentioned before, the string passes over the bridge so that when the

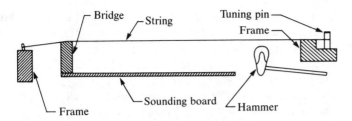

Figure 4.2. A string, the bridge, and the sounding board in a piano

string vibrates, the bridge does also. The bridge, in turn, transfers these vibrations to the sounding board. The large area of the sounding board transfers the vibrations to the air effectively. In the language of Chapters 2 and 3, the sounding board couples to the air much better than the string does. Also, the sounding board is an oscillator that is driven by the vibrating string. The response curve for a good sounding board is very different from the response curves we already have described, however. This fact becomes obvious when we list the desirable characteristics of a sounding board.

1. The board should respond equally well to the full range of frequencies at which the piano's strings vibrate. Otherwise, some notes would sound unusually loud or soft.

2. The vibrations of the sounding board itself must decay very rapidly when the driving vibration is removed. As soon as the pianist stops the vibration of a string (by releasing the key), the sound should stop, and this "quick reaction" will happen only if the sounding board has a short damping time.

In order to meet these requirements, a sounding board must have an extremely broad response curve that will be flat on top.

Although the sounding board provides a much better transfer of a string's vibration to the air, the sound of one string still would be too soft for all but the lower notes on the piano. To overcome this difficulty, pianos have multiple strings for each of the higher notes. The top sixty-eight notes have three strings each; the next twelve notes have two strings each, while the lowest eight notes have only one string each.

For a note having multiple strings, each string is tuned to a slightly different frequency. This setup gives a much richer or more interesting tone quality. Some scientists believe the beats that occur between all the respective normal modes of the different strings produce this enhanced tone quality.

In the piano, a hammer hits a string to start the string vibrating, and a separate device called a *damper* is used to stop the vibration. While a note is sounding, of course, the damper must be lifted off the string. The movement of both the hammer and damper for each string is controlled by the pianist, who presses the corresponding piano key. The complicated mechanism shown in Figure 4.3 transforms the piano key's movement into the correct motions of the hammer and damper. This mechanism, called the *action,* is complicated because it must do so many things. First, the player must be able to control how hard the hammer strikes the string. Then, when the hammer does hit the string, the hammer must jump away immediately without bouncing back against the string. The mechanism also must lift the damper off the string just as the hammer hits the string and then must replace the damper on the string when the key is released. Finally, the player may want to repeat a note rapidly, so the mechanism must be able to do this. The actual operation of the mechanism shown in Figure 4.3 is not important for our purposes; the fact that the mechanism performs the tasks just described is sufficient.

A couple of details about the hammer and damper themselves are interesting, however. Each of these devices is covered with thick felt, and the actual hardness of this felt affects the vibration of the string. A hard surface on the hammer felt enhances the higher overtones in the string's vibration recipe. When a tone's vibration recipe has large amounts of higher overtones, musicians say the tone has a "brighter" sound. Conversely, a soft felt on the hammer favors the lower normal modes in the vibration recipe, giving a "mellow" tone. The condition of the felt on the damper determines how rapidly the note is cut off. A note sounds different when it is stopped abruptly, compared to when it takes several cycles to fully stop. The surface of the felt

Figure 4.3.   The action of a piano

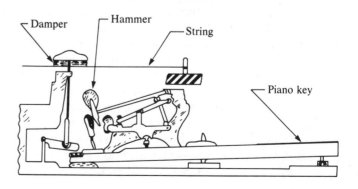

can be adjusted by a piano tuner. Pricking the surface with a pin will soften the felt, and heating the surface with a hot iron will harden the felt.

Pianos have either two or three pedals that allow the player to change the behavior of the instrument. The pedal on the player's right lifts the dampers from all the strings. This complete lifting causes each note played to continue sounding even after the key has been released. As a result, the piano sounds louder, so we call this pedal the *loud pedal*. The pedal on the player's left is called the *soft pedal* because depressing it causes a softer sound. Piano builders use different mechanisms to produce this effect. In one way (which often is used in upright pianos), the pedal physically moves the hammers closer to the strings. Then, when the player hits the key causing hammer movement, the hammer does not move as far, so the hammer is not traveling as fast when it strikes the string. In another method (which is used in grand pianos), the hammers move to one side. For those notes having multiple strings, this sideways displacement causes the hammer to hit only one of the strings. With the first method, each string has a smaller vibrational amplitude, while in the second method, fewer strings vibrate, but with the same amplitude. The final result is the same: The sounding board has a smaller amplitude, so the sound is softer. If there is a third pedal, it will be between the previous two. When depressed, this pedal causes the dampers for the lower notes that are already raised to remain off their strings. This allows the performer to sustain a chord while playing other notes.

Now, let's consider the strings themselves. Piano strings are real strings in contrast to the ideal, perfectly flexible ones considered in Chapter 3. Thus, there are some differences in the way piano strings vibrate. The most notable difference is that the normal modes of a piano string are not harmonics. Recall that when we have harmonics, the frequency of each overtone is an integral multiple of the fundamental frequency. In particular, the second normal mode has a frequency that is twice that of the first normal mode. Because the piano string is stiffer than an ideal string, the frequency of its second normal mode is slightly higher than twice the frequency of its fundamental. The same thing is true for all the other higher modes of vibration for a piano string. This situation is one of the reasons for the piano's distinctive sound. Scientists have been able to reconstruct the vibration recipe of a piano note using electronic instruments. The scientists then have changed the frequencies of the overtones to obtain harmonics. But when this is done, the resulting

The action of a grand piano (Picture provided courtesy of the Kimball Keyboard Division.)

The strings of a grand piano with the damper for one string being lifted (Picture provided courtesy of the Kimball Keyboard Division.)

tone no longer sounds like a piano! This deviation from the harmonic structure of the overtones also can cause an interesting situation. One way a piano tuner performs his task is by matching the fundamental frequencies of high notes to overtone frequencies of low notes. Since the overtone frequencies are higher than harmonic frequencies, this method produces a *stretched* musical scale.

Although people accustomed to hearing a piano can detect this inharmonicity, the actual differences are small. The peaks on the response curves for strings that are an even number of octaves apart will overlap. You easily can demonstrate this concept to yourself with the following simple experiment: First, depress a key on the piano, but do it slowly enough so that the hammer does not strike the string. (By holding this key down, you will hold the damper off its string.) Now, play the note one octave lower; hold it for a moment, and then, release it. After the lower note ceases, you should hear the higher one until you release that key. This is an example of resonance between the first overtone of the lower frequency string and the fundamental of the higher frequency string. Note that resonance occurs even though the two frequencies do not match exactly; only portions of the respective response curve peaks overlap. Because the difference between the normal mode frequencies of a real piano string and the harmonics of an ideal string is small, we usually will not consider it in our discussions.

A very distinctive aspect of a piano's sound is the *attack*. This term refers to how the note begins. On the piano, of course, the note begins when the hammer strikes the string. When this striking happens, many modes of the string are set into vibration. This full vibration recipe produces a very rich tonal quality. The vibration recipe changes with time, however, because the different modes of vibration have various damping times. Generally, higher modes of vibration decay more rapidly than lower modes. Thus, after the piano key for a low-pitched tone is held depressed for several seconds, the only significant mode of vibration remaining will be the fundamental. This situation occurs more rapidly for higher-pitched tones; the waiting period is only fractions of a second for the tones in the highest octave.

We can observe the changing vibration recipe of a piano note in several ways. First, we can use a strobe and observe the vibrating string directly. We set the frequency of the strobe equal to the frequency of the string's fundamental. Immediately after the string is struck with the hammer, the string's motion will appear to be very irregular. When the higher modes have damped out, however, and the string is vibrating in its fundamental mode, we will see that the string's motion is very regular. Another way to observe the changing vibration recipe is to use an oscilloscope to see the vibration pattern of the sound that is produced. At first, the vibration will be very complicated, but it will smooth out as the higher modes damp away. We can refine this method by adding an electronic filter. This filter allows only frequencies that fall within a specified range to be observed on the oscilloscope screen. By changing the range, we can observe in turn the vibration of each normal mode of the string. With this equipment, we could determine each normal mode's damping time. Usually (but not always), we would find that successively higher modes have shorter damping times.

Thus, we see that many factors enter into the production of vibrations that cause the sound we recognize as piano music. Scientists have used various techniques to alter these vibrations by changing the vibration recipe, inharmonicity, attack, or decay. A person with musical training easily recognizes that the resulting sound is not coming from a piano. There seems to be no fundamental reason why piano music sounds pleasant. Apparently, one's hearing becomes conditioned to the unique aspects of piano sounds, and any variation is disconcerting.

Musical Scales

As noted earlier, the fundamental division in any musical scale is the octave. Moving up the scale by one octave always doubles the frequency. Thus, if we start with a note having a frequency of 50 Hz, the note one octave higher will have a frequency of 100 Hz. Likewise, the note another octave higher has a frequency of 200 Hz. Notice that we are concerned only with the frequency ratios, not with frequency differences. (In each case, the frequency ratio is 2:1; in the first case, the frequency difference is 50 Hz, and in the second case, the difference is 100 Hz.) The name octave comes from the fact that on the piano, we can start with any white key, label it number one, and label successive white keys two, three, four, etc. Then, the note one octave higher than key number one is key number eight, and octave means eight in Latin.

Similar names are used for other intervals between notes or keys on the piano. These relationships are understood most easily when we start numbering on the note *C*, because we then can count using only the white keys. Two intervals will be important in our discussion: the *fourth* (keys one and four, or notes *C* and

F), which has a frequency ratio of 4:3; and the *fifth* (keys one and five, or notes *C* and *G*), which has a frequency ratio of 3:2. In the modern scale, each octave is divided up into a set of twelve notes, but musical scales have not always had this construction. Before discussing the modern musical scale, we briefly will review the historical development of musical scales.

When two musical tones are heard together, they can sound pleasant or unpleasant. In the former case, we term the tones *consonant*; while, in the latter case, we term them *dissonant*. Usually in a musical performance, the composer and performers want to create notes that are consonant when sounded together. (Occasionally, for special effect, dissonant notes are desirable.) In ancient Greece, the School of Natural Philosophy, founded by Pythagoras, studied the question of consonance and dissonance. The members of the school concluded that two notes whose frequencies are ratios of small integers will sound consonant. Note that the intervals of the octave, the fourth, and the fifth satisfy very well this condition for consonance established by the Pythagorean school.

Using these intervals to subdivide each octave, a musical scale was developed, which is now termed the *Pythagorean* scale. This scale is theoretically attractive, but it has certain practical limitations. To demonstrate its theoretical appeal, we will look at how this scale is developed. Referring to Table 1, we start with a *key note* or *tonic,* which we label note 1, and (for ease of calculation) we assign it a frequency of 100 Hz. (Rather than using letters for the notes in this scale, I am using numbers because the scale is only hypothetical.) The note one octave above note 1 (which we call 1′) has a frequency of 200 Hz. Now, we can add notes 4 and 5 with known frequencies using the frequency ratios for the fourth and fifth. Note 4 has a frequency of 133 Hz, which is found by multiplying 100 Hz by 4/3. Likewise, note 5's frequency is 150 Hz.

To get the other notes in the scale, we must be a little more clever. We start at note 5 and go up another fifth, which will give note 9 with a frequency of 225 Hz (150 multiplied by 3/2). But note 9 is in the next octave, so we rename it 2′ and drop down an octave getting note 2 with a frequency of 112.5 Hz. (The ratio of this frequency to that of note 1 is 9:8, so it is close to satisfying the Pythagorean condition.) To get note number 6, we go up a fifth from note 2. Thus, the frequency of note number 6 is 112.5 × (3/2) = 168.75 Hz, and the ratio of this frequency to that of note 1 is 27:16. Thus, in this scale, the Pythagorean school's condition for consonance is not satisfied very well for notes 1 and 6, and the musical sixth presumably would not sound nice.

Table 1.   The construction of a Pythagorean scale

Note	Frequency (Hz)	Frequency Ratio to Note One	Frequency Ratio to Next Lower Note
1	100.	1:1	—
2	112.5	9:8	9:8
3	126.56	81:64 [9:8 with note two]	9:8
4	133.33	4:3	256:243
5	150.	3:2	9:8
6	168.75	27:16 [3:2 with note two]	9:8
7	189.84	243:128 [9:8 with note six]	9:8
8 or 1′	200.	2:1	256:243
9 or 2′	225.	[3:2 with note five]	9:8

These five notes (numbers 1, 2, 4, 5, and 6) form a five-note or *pentatonic* scale. This scale has been developed using only intervals of the fourth, the fifth, and the octave and is common to many cultures. Music having only these notes will sound oriental or eastern to most of us who are used to music from western culture.

Two more notes are needed to complete the Pythagorean scale. To get each of them, we use the frequency ratio between notes one and two; that is 9:8. Applying this ratio to note two gives note three with a frequency of $112.5 \times (9/8) = 126.56$ Hz. Note seven is found in the same way from note six and has a frequency of $168.75 \times (9/8) = 189.84$ Hz.

In this scale, the frequency ratio for adjacent notes is shown in the fourth column of Table 1. The ratio 9:8 is larger than the ratio 256:243; the first interval is termed a *Pythagorean whole tone,* and the second is called a *Pythagorean semitone.* These intervals fit the white keys on the piano (when note number one is a *C*), so the piano could be tuned with these frequency ratios.

Whenever an interval is made with two notes having frequencies that are the ratios of small whole numbers, we call it a *just* interval. (In this usage, just means good, as in justice.) Thus, the fourth (ratio of 4:3) and the fifth (ratio of 3:2) are just intervals. Note that the Pythagorean scale does not contain any just thirds or sixths, only fourths and fifths. To overcome this apparent defect, a new scale was developed by musical theorists. Because it has many pairs of notes forming just intervals, it is called the *just diatonic* or *natural diatonic* scale and is displayed in Table 2. The number names of the notes are in column one, and their frequencies are in column two. Column three lists the ratio between the frequency of the note and note 1. It is quite evident that we have achieved a just third (ratio of 5:4), a just fourth (ratio of 4:3), a just fifth (ratio of 3:2), and a just sixth (ratio of 5:3). However, we also have lost something! In the Pythagorean scale, the frequency ratio of note 6 to note 2 was a just fifth, but here, the ratio is 40:27, which cannot be considered a just interval. Further investigation (which will not be developed here) shows this to be a fundamental problem. That is, it is impossible to write down a set of frequencies within an octave for which all intervals are just. Thus, musicians have compromised the ideal of just intervals and have dropped the Pythagorean and just scales.

The scales in Table 2, developed to use just intervals, have been replaced with the *equally tempered* scale. In this scale, the twelve notes within an octave (this includes the seven white keys and five black keys on a piano) are designed so the same interval occurs between each adjacent pair. That is, the frequency ratio between any two adjacent notes is the same. We call this frequency ratio between adjacent notes r. And $r = f_2/f_1 = f_3/f_2 = f_4/f_3$, etc. Since there are twelve separate notes within each octave, there will be twelve such intervals. We know the frequency doubles for each octave, so we must have $r^{12} = 2$. (The left-hand side of this relation is the twelve steps, each of size r, and the right side is the factor of two for an octave.) This correspondence, of course, means that the frequency ratio between any two adjacent notes in this scale is

$$f_2/f_1 = r = \sqrt[12]{2} = 1.059463 \qquad \textbf{(4.1)}$$

Table 2.   The construction of a just diatonic scale

Note	Frequency (Hz)	Frequency Ratio to Note One	Frequency Ratio to Next Lower Note
1	100.	1:1	—
2	112.5	9:8	9:8
3	125.	5:4	10:9
4	133.33	4:3	16:15
5	150.	3:2	9:8
6	166.67	5:3	10:9
7	187.5	15:8	9:8
8 or 1′	200.	2:1	16:15

Table 3. The construction of an equally tempered scale

Note	Frequency (Hz)	Frequency Ratio to Note 1*	Frequencies of Added Notes (Hz)
1	100.	1	
		r	105.95
2	112.25	r^2	
		r^3	118.92
3	125.99	r^4	
4	133.48	r^5	
		r^6	141.42
5	149.83	r^7	
		r^8	158.74
6	168.18	r^9	
		r^{10}	178.18
7	188.77	r^{11}	
8 or 1'	200.	r^{12}	

*$r = \sqrt[12]{2} = 1.0594631$

In Table 3, we have our hypothetical scale (which is now equally tempered) starting with 100 Hz. We easily can compare frequencies in the three different scales to see what shifts were made for various notes in going from one scale to another. It is important to recognize that note 2 is two intervals above note 1. We call this interval a *tempered whole tone*. The same thing is true of the interval between notes 2 and 3. The interval between notes 3 and 4, however, is a *tempered semitone*, and the ratio of their frequencies is r. The last column in Table 3 shows the frequencies of those notes in this scale that correspond to the black notes on the piano.

Let's compare the frequencies of notes in the presumably desirable, just diatonic scale and the pragmatic, equally tempered scale. The biggest change has occurred for note 6, where the frequency has gone from 166.67 Hz to 168.18 Hz. However, this latter value is closer to that occurring in the Pythagorean scale, so it is a good compromise between the first two scales. It turns out that only people who have a very good musical ear can detect the difference between the just scale and the equally tempered scale. It also seems that people like the scale they are used to hearing. Since most music is played using the equally tempered scale, music may sound strange when played using a just scale.

Finally, Table 4 displays the frequencies for each tone in the current widely used version of the equally tempered scale. The table also shows a standard notation for notes where their letter names are given subscripts to indicate the octave to which they belong. *Middle C,* which was mentioned earlier in this chapter, is C_4 with a frequency of 261.63 Hz, and the A above *Middle C*

Table 4.   Frequencies of notes in the equally tempered scale

Note	Frequency (Hz)	Note	Frequency (Hz)	Note	Frequency (Hz)
C_0	16.352	C_3	130.81	C_6	1046.5
	17.324		138.59		1108.7
D_0	18.354	D_3	146.83	D_6	1174.7
	19.445		155.56		1244.5
E_0	20.602	E_3	164.81	E_6	1318.5
F_0	21.827	F_3	174.61	F_6	1396.9
	23.125		185.00		1480.0
G_0	24.500	G_3	196.00	G_6	1568.0
	25.957		207.65		1661.2
A_0	27.500	A_3	220.00	A_6	1760.0
	29.135		233.08		1864.7
B_0	30.868	B_3	246.94	B_6	1975.5
C_1	32.703	C_4	261.63	C_7	2093.0
	34.648		277.18		2217.5
D_1	36.708	D_4	293.66	D_7	2349.3
	38.891		311.13		2489.0
E_1	41.203	E_4	329.63	E_7	2637.0
F_1	43.654	F_4	349.23	F_7	2793.8
	46.249		369.99		2960.0
G_1	48.999	G_4	392.00	G_7	3136.0
	51.913		415.30		3322.4
A_1	55.000	A_4	440.00	A_7	3520.0
	58.270		466.16		3729.3
B_1	61.735	B_4	493.88	B_7	3951.1
C_2	65.406	C_5	523.25	C_8	4186.0
	69.296		554.37		4434.9
D_2	73.416	D_5	587.33	D_8	4698.6
	77.782		622.25		4978.0
E_2	82.407	E_5	659.26	E_8	5274.0
F_2	87.307	F_5	698.46	F_8	5587.7
	92.499		739.99		5919.9
G_2	97.999	G_5	783.99	G_8	6271.9
	103.83		830.61		6644.9
A_2	110.00	A_5	880.00	A_8	7040.0
	116.54		932.33		7458.6
B_2	123.47	B_5	987.77	B_8	7902.1

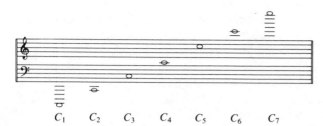

Figure 4.4. Various C notes in place on the musical staves

(A_4) has a frequency of 440 Hz. This latter note is the standard from which all the other frequencies are derived, using the ratio r an appropriate number of times. Although this musical scale currently is accepted and used widely throughout the world, it is not universal. Some orchestras will tune A_4 to 442 or 444 Hz and adjust the frequencies of all the other notes accordingly.

As mentioned earlier, the piano is a fixed-frequency instrument. That is, the player cannot adjust the frequency of the notes while playing it. This condition is not true for many other instruments such as members of the violin family, the human voice, and the trombone. The flexibility of these variable-frequency instruments means they can be played using notes in any musical scale. The fixed-frequency instruments (such as the piano, woodwinds, guitars, harp, and harpsichord) must be tuned to a specific scale. The equally tempered scale was developed as a compromise to accommodate fixed-frequency instruments. The notes of this scale are placed on musical staves for easy reading. Figure 4.4 shows seven different C's placed on these staves, and it is easy to see why *middle C* (C_4) gets its name.

In later chapters, we will discuss several other instruments. We will find that the vibration recipe, attack, and decay of the notes played give each instrument its distinctive sound. The design of each instrument and the way each instrument is played determine these aspects of its sound. Before we look at these other instruments, however, we first will investigate the hearing process and consider some information about waves.

Summary

The piano is a complicated instrument, and the string is its basic vibrating system. The length, tension, and weight of the string determine the frequencies of the string's normal modes. The string's vibrations are transferred through the bridge to the sounding board; the sounding board is used to transfer the vibrations to the air more effectively. Many different components within the piano help to produce its distinctive sound: Strings, dampers, hammers, the sounding board, and the action all contribute.

The fundamental division of any musical scale is the octave; in the modern scale, each octave is divided into a set of twelve notes. The equally tempered scale was developed to accommodate fixed frequency instruments like the piano. In this scale, the twelve notes within an octave are designed so the frequency ratio between two adjacent notes in the scale is the same: $f_2/f_1 = r = \sqrt[12]{2} = 1.059463$.

Suggested Readings

The article, "The Physics of the Piano," by E. Donnell Blackhaus (*Scientific American.* December 1965.) describes the piano's physical structure, action, and change in vibration recipes for various notes with time.

Both the piano and musical scales are discussed in *Physics and Music,* by Harvey E. White, and Donald H. White (Philadelphia: Saunders College/Holt, Rinehart, and Winston, 1980). The section on the piano is rather brief, but the discussion of musical scales is detailed.

A much more complete discussion of the piano is contained in *The Piano—Its Acoustics,* by W. V. McFerrin (Boston: Tuners Supply Co., 1972). The book contains a great deal of detail (almost too much), particularly about the construction of the piano. However, the writing has no difficult scientific or mathematical concepts to worry about.

Review Questions

1. A piano tuner notes that the tone produced by a string is too low. In order to adjust the string to produce the proper tone, he

   a. wraps another piece of wire around it to increase its weight.

   b. turns the tuning pin to increase the tension.

   c. turns the tuning pin to decrease the tension.

   d. moves the tuning pin so the string becomes longer.

2. Piano strings differ from ideal strings because

a. they are made of wire and are affected by the magnetic field of the earth.

b. they are stiffer, so the higher harmonics are stretched, producing wider peaks on the response curve.

c. they are stiffer, so the overtones have higher frequencies than harmonics would have.

d. they are heavier, so the peaks on the response curve for the overtones are stretched into much taller peaks.

3. The strings that produce the low tones on the piano are wrapped because

a. wrapped strings resonate better with the sounding board.

b. wrapped strings have two overtones within each octave.

c. wrapped strings have lower frequencies than unwrapped strings of the same length and tension.

d. wrapped strings have overtones that are closer to harmonics than unwrapped strings.

e. wrapped strings are easier to play.

4. The bridge in the piano

a. transfers the motion of the key to the hammer.

b. transfers the motion of the string to the damper.

c. transfers the motion of the string to the sounding board.

d. transfers the motion of the hammer to the string.

e. transfers the motion of the sounding board to the frame.

5. Which one of the response curves in Figure 4.5 reflects a good piano sounding board?

a. A

b. B

c. C

d. D

e. E

Figure 4.5. For use with review question 5

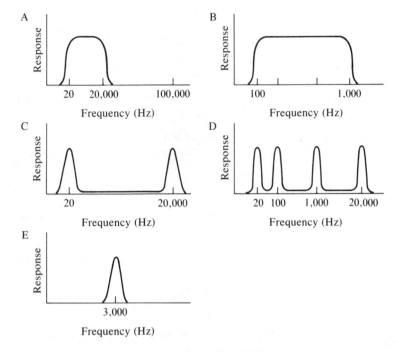

6. The key for note A_3 (220 Hz) is held down, and the key for A_1 (55 Hz) is struck and released. The tones heard will have frequencies of, respectively,

   a. 55 Hz and 110 Hz.

   b. 55 Hz, 110 Hz, and 220 Hz.

   c. 55 Hz and 220 Hz.

   d. 110 Hz and 220 Hz.

   e. 220 Hz only.

7. Which of the following is *not* a fixed-frequency instrument?

   a. Trombone

   b. Clarinet

   c. Harp

   d. Piano

   e. Flute

8. Which of the following is a fixed-frequency instrument?

   a. Violin

b. Saxophone

c. Trombone

d. The human voice

e. Cello

9. The primary advantage of the natural diatonic scale is that

   a. it can be played on all types of instruments.

   b. the overtones of the notes are stretched, so they match those overtones that occur on the piano.

   c. the ratios of all other frequencies in the scale to that of the tonic are proportional to ratios of small numbers.

   d. the ratio of frequencies of adjacent tones always has the same value.

10. One tone is produced with a frequency of 150 Hz. The tone three octaves higher is produced with a frequency of

    a. 300 Hz.

    b. 450 Hz.

    c. 600 Hz.

    d. 900 Hz.

    e. 1200 Hz.

5

The Ear as a Detector of Sound

Important Concepts

- An average young, normal ear has
  1. an audible frequency range from 20 Hz to 20,000 Hz.

  2. a maximum sensitivity to sound with frequencies around 3000 Hz.

  3. an ability to hear sound intensities as low 10^{-12} watts per square meter.

  4. an upper limit for the intensity of sound—as determined by the onset of pain—of about $0.1 - 1.0$ watts per square meter.

- The psychological variables of pitch, loudness, and tone quality (or timbre) depend primarily on the physical variables of frequency, intensity, and vibration recipe, respectively.

- The ear can detect frequency changes of 3 Hz below 1000 Hz, and it approximately follows Weber's law above 100 Hz, being sensitive to a change of 0.3 percent in frequency.

- The ear is sensitive to changes in intensity of about 6 percent for loud sounds, but for soft sounds, the change must be about one hundred percent before the ear can detect it.

- Difference tones are produced in the same way as beats; thus, the frequency of a difference tone is equal to the difference of the frequencies producing it.

- A tone can mask another of higher frequency more easily than one of lower frequency.

- Sound vibrations in the ear are channeled by the auditory canal to the eardrum. This passage has a resonant frequency of about 3000 Hz, making the ear most sensitive to that frequency.

- Mechanical vibrations are passed from the eardrum through the bones in the middle ear (the hammer, anvil, and stirrup) to the oval window and the fluid in the cochlea.

- The vibrating fluid in the cochlea drives hairs mounted in the organ of Corti on the basilar membrane. That motion stimulates the auditory nerve endings located in the organ of Corti, sending nerve signals to the brain.

- According to the place theory of hearing, pitch is determined by the location on the basilar membrane having the largest vibrational amplitude, and loudness is determined by the extent to which the membrane's length is vibrating.

- The production of combination tones in the ear provides evidence that the ear is a very complicated vibrating mechanism.

In this chapter, we will investigate the response of the ear to sound vibrations. First, we will discuss the sounds that a normal ear can detect to see how well it works. Then, we will investigate the ear's physiological structure to understand the way it does the job.

Capability of the Ear

The ear is a wonderful instrument for detecting sound. Its great capability stands out when we explore the ranges of frequency and intensity of sound it can detect. For a young person with normal hearing, the frequency of sound can be varied by a factor of almost a thousand and still can be heard; the nominal range of audible frequencies is from 20 Hz to 20,000 Hz. The intensity range seems even more remarkable. Very loud sounds with an intensity of about 1 watt per square meter (W/m^2) are at the threshold of producing pain. At the other end of the intensity spectrum, we have barely audible sounds such as a soft whisper or the wind rustling leaves on a tree. These sounds have an intensity of about 10^{-12} watts per square meter, which is one trillionth as large as the painful sounds. Thus, a young person with normal hearing can detect sounds that differ by a factor of a thousand in frequency and by a factor of a trillion in intensity.

No two people have exactly the same capabilities for detecting sound. There are three basic reasons for these differences: the construction of the ear (which is determined by heredity), the age of the person, and the history of the ear. We will discuss shortly the construction of a typical ear. During that discussion, you should keep in mind that small differences in the size or shape of some part of the ear will produce differences in hearing ability. The age of the ear also affects capability because when we grow old, we cannot perform as well physically as we could during our teens and twenties. In support of this statement, note that Olympic records are set by young men and women. The ear is a physical mechanism; it has parts that move much like a mass on the end of a spring. As we grow older, not only do muscles become weaker, but other tissues such as tendons and cartilage also do not function as well as they once did. With the ear, our ability to hear high frequency sounds deteriorates first. A person with normal hearing finds his upper limit of frequency has declined to about 14,000 Hz by the time he reaches age forty and continues to decline as he grows older. The history of the ear is important because very loud sounds damage the delicate mechanisms in the ear. There are members of a native tribe in northern Africa who are in their fifties but still can hear as well as young people. They live in a quiet society! In contrast, people who live in a noisy world lose their ability to hear. The source of the noise does not matter. The noise can be produced by machinery, jet planes, or electric guitars. An extended dose of loud sounds will produce a severe deterioration in a person's ability to hear.

In discussing the performance of the ear, we must recognize there are two ways to describe the sounds we hear. First, there is the objective description, which physicists use and which includes the terms *frequency*, *intensity*, and *vibration recipe*. This set of descriptors is objective because the descriptors do not depend upon someone's interpretation of the sound; they can be measured with instruments. Sound also can be described subjectively using the terms *pitch*, *loudness*, and *tonal quality* or *timbre*. These descriptors cannot be measured with instruments;

their values are determined by someone's interpretation. Although there is a correspondence between these two ways of describing sound, they are not just directly related to one another. Instead, there is a strong direct relationship and a small cross relationship. This dual correspondence is illustrated in Figure 5.1; the heavy arrows denote the primary relationship between these ways of describing sound, and the lighter arrows indicate secondary relationships. Now, we will discuss these interrelationships.

First, we will talk about the pitch of a musical note. If we use an electronic signal generator to produce a pure tone, then the frequency and amplitude of the tone can be varied easily. In Chapter 1, we discussed this experiment, recognizing that increasing the frequency caused an increase in pitch while increasing the amplitude caused a louder tone. Now, however, we want to consider a complicated system that is vibrating in several modes at the same time. In this situation, all these vibrations of different frequencies arrive at the eardrum simultaneously. When the frequencies are harmonic (or nearly harmonic, as for a piano), the ear picks out the fundamental frequency from all these vibrations and uses it to specify the pitch. Even though the vibration recipe is changed to one that is very different, if the fundamental frequency stays the same, the pitch of the tone does not change. (This statement covers most situations, but later in this chapter, we will look at some special cases that are exceptions to this rule.) Thus, when we hear notes of the same pitch played by a violin, a trumpet, and a clarinet, we know the fundamental frequencies of these three notes are the same. However, the notes sound different. Many people could tell which notes came from which instrument. That identification is based partly on the tonal quality of the notes. (Other factors that help to distinguish the tones played by various instruments include the *attack* and *decay* of the sound. The attack is the way the note starts, and the decay is the way it ends.) As we will discuss shortly, the tonal quality is most closely associated with the vibration recipe of the sound.

Figure 5.1. Comparing the two ways of describing sound

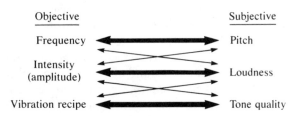

As we have seen, the source of sound is vibration. When the amplitude of the vibration gets larger, the sound gets louder. But we can be more precise in this description. First, we introduce the *intensity*, which is proportional to the square of the vibration amplitude. Thus, if the vibration amplitude is doubled, then the intensity increases by a factor of four. Loudness is most directly related to the intensity, but this relationship is complicated for two reasons. First, different people perceive the same sounds differently. More important for our purposes, however, are the different responses an average ear has to different sounds.

The way a person with average hearing perceives pure tones of different frequencies and intensities is shown in Figure 5.2. The numbers across the bottom scale show the frequency of the vibration. The numbers along the left-hand side show the intensity of the sound. You should recognize that objective, physical variables are plotted along both axes of this graph. In contrast, the wiggly lines across the graph are lines of equal loudness (for an average person), and this loudness is a subjective, psychological variable. Thus, this graph connects the subjective variable loudness to the objective variables of frequency and intensity. Sounds lying on the lowest wiggly line running across the graph are barely perceptible, so we call this line the threshold of hearing. Each successively higher line is a louder sound, and sounds lying along the top line are actually painful because they are so loud.

We can use this graph to compare the intensities needed to make sounds appear equally loud when the frequencies of these

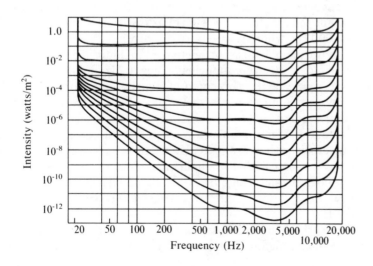

Figure 5.2. Loudness as it depends on frequency

sounds are different. For example, suppose we have two pure tones with frequencies of 100 Hz and 1000 Hz. Each of these two tones will be barely perceptible to a person with average hearing if the low frequency tone has an intensity of $8. \times 10^{-9}$ watts per square meter and the high frequency tone has an intensity of $1. \times 10^{-12}$ watts per square meter. Thus, both tones sound equally loud (we just barely can hear them), but their intensities differ by a factor of about 8000.

Because the graph tells us about the ability of a person with average hearing to perceive sound, the graph describes the sensitivity of the average ear. In doing so, the graph tells us about many facets of human hearing. First, we note that humans cannot hear either very low or very high frequency sounds. No matter how high the intensity gets at these frequencies, the threshold of hearing is not crossed. Second, we note that a person's ability to hear low intensity vibrations varies greatly with frequency. The minimum in the threshold of hearing curve is about 3000 Hz; at this point, the ear is most sensitive. At other frequencies, the intensity of the sound must be greater in order for the ear to detect it. Finally, we note that for very high intensity vibrations (that is, loud sounds), little difference exists in the ear's sensitivity throughout the audible range of frequencies.

These curves of equal loudness give the same information as the response curve of an oscillator. (In this case, the ear is the oscillator, and the curves of equal loudness show how well it can be driven at various frequencies.) Compared to a response curve, however, these curves are upside down. Let me clarify this statement. We know that energy is transferred to an oscillator most easily at the frequency where the oscillator's response curve peaks. For the ear, however, energy is transferred most easily at about 3000 Hz, where these curves have dips. Thus, these curves are like upside-down response curves. In addition, the ear does not respond at all to frequencies below 20 Hz and above 20,000 Hz. The rising of the threshold of hearing curve to very high levels is the same as a response curve being zero in this frequency range.

Note that Figure 5.2 shows equal loudness curves for a young person with average hearing. These curves could be very different for someone with a hearing deficiency. As an example, for an older person, the curves all would rise steeply at a frequency much lower than that illustrated (which occurs at 20,000 Hz).

From the above discussion, we see that the loudness of a sound depends primarily on its intensity, but it also depends upon the frequency of the vibration. As we will see shortly, this behavior is caused by the structure of the ear.

For some people, an effect occurs that is opposite to the one just described. That is, when the intensity of a vibration is changed rapidly, such people perceive a change in pitch even though the frequency remains fixed. This effect is much less noticeable, and it occurs for only a limited range of frequencies. Including it with the above discussion on how loudness varies with frequency, the effect shows us there are crossover effects in both directions. It is certainly true, however, that pitch depends primarily on frequency and loudness depends primarily on intensity.

In unusual cases, pitch also can depend on the vibration recipe. In order to explain how this dependence can occur, I first need to describe difference tones. As we saw in Chapter 2, two vibrations of different frequencies occurring at the same time will produce beats. And the beat frequency is equal to the difference in frequency of the two vibrations. In the case of sound, we saw that a 300 Hz tone and a 320 Hz tone will produce a 310 Hz tone with a variation in amplitude (or intensity or loudness) at a rate of 20 Hz. Now, suppose the 300 Hz tone is held fixed and the frequency of the 320 Hz tone is increased steadily. Then, the rate of beats (or the beat frequency) will increase also. When this rate is high enough (at least greater than 20 Hz), it produces a vibration of the eardrum that is within the audible range. Thus, we detect the sound as a note of a new frequency—a frequency that is equal to the difference of the two frequencies producing it. Thus, we call it a *difference tone*.

Now, let's see how the existence of difference tones can help us understand a case in which the pitch of a note is not determined by the lowest frequency present in the vibration recipe. Suppose we play a note with a musical instrument for which the modes of vibrations are harmonics. For a fundamental frequency of 200 Hz, there also will be vibrations at 400 Hz, 600 Hz, 800 Hz, etc. Between each pair of adjacent normal modes, there will be beats—that is, difference tones—of frequency 200 Hz. Electronically filtering the note to remove the fundamental frequency will not change the pitch of the tone. This condition is true even though the pitch usually is determined by the lowest frequency present. Since the fundamental frequency of 200 Hz has been filtered out, we would expect the pitch to rise so that the pitch corresponds to a frequency of 400 Hz. The reason there is no change, of course, is because our ears still hear the fundamental frequency, except that now, it is a difference tone. Thus, the vibration recipe primarily determines the tonal quality or timbre of a sound. But in unusual cases, the vibration recipe can affect the pitch also.

In addition to the frequency and intensity ranges that the human ear can detect, it is also interesting to look at the ear's ability to tell when two nearly identical pure tones are different. Suppose we start with two identical pure tones (they have exactly the same frequency and intensity), and we change one of the tones, either by changing its frequency or its intensity. The question we are interested in here is: How big a change is needed in order for a normal ear to know that a change has been made? This ability is known as *differential pitch discrimination* (in reference to frequency changes) and *differential loudness discrimination* (in reference to intensity changes).

In 1834, Ernst H. Weber, a German physicist, suggested what has become known as Weber's Law: A stimulus must increase by a constant fraction of its value in order to be noticeably different. This statement means that if we have to change the frequency of a note from 100 Hz to 102 Hz in order to detect the change, then the frequency of a 1000 Hz note would have to change to 1020 Hz to be detected. (Note that the frequency change divided by the original frequency is the same in each case.) Although this law is only approximately true, it serves as a good starting point. For example, in pitch discrimination, a normal ear can detect a change of about 3 Hz for any frequency below 1000 Hz (which does not follow Weber's Law). But above 1000 Hz, the ratio of the frequency change to the frequency itself that a normal ear can detect stays about the same, that is, a 0.3 percent change. Thus, for example, when a pure tone of 1200 Hz is changed to 1236 Hz, most people will notice a difference in pitch. (This is a 0.3 percent change.) Similar behavior occurs with loudness discrimination. Weber's Law is again only approximately true, but the deviations from it in this case are more complicated; they depend on both intensity and frequency. A reasonable, approximate relation is that a typical ear can tell when the intensity of a sound changes by about 6 percent for loud sounds but requires an almost one hundred percent change for faint, barely perceptible sounds.

One final effect I wish to describe is *masking*, which occurs when two tones are sounded together and one of them drowns out the other. We say the predominant tone masks the one that cannot be heard. The interesting fact about masking is that a low frequency tone can mask a high frequency one much more easily that the other way around. To explain what I mean by this, consider what happens when the tones of 300 Hz and 1300 Hz are sounded together. If the two tones have low intensities and about equal loudness, then both tones can be heard. As a specific example, let's set the intensity of the 1300 Hz tone at 10^{-11} watts per square meter and that of the 300 Hz tone at 3×10^{-10} watts per square meter. From Figure 5.2, we see that these two soft tones will sound equally loud to the average ear. We can increase the intensity of each tone separately. We find that when either one is increased (and the other is kept at the original low intensity), then the louder one eventually drowns out (or masks) the softer one. Masking by the 300 Hz tone occurs when its intensity is increased to about 10^{-7} watts per square meter. On the other hand, even when the intensity of the 1300 Hz tone is 10,000 times bigger than this (that is, an intensity of 10^{-3} watts per square meter), it has difficulty masking the original 300 Hz tone. This example shows what we mean when we say it is easier for a low frequency tone to mask a higher frequency one than vice versa.

Structure of the Ear

Now let's turn to the structure of the ear and see why it performs the way we have described. Figure 5.3 shows a cross section through the ear. The cross section contains all the parts of the ear required for the detection of sound. First, note that the ear divides very naturally into three parts: the outer, middle, and inner ears. The outer ear (which is the only visible part) ends at

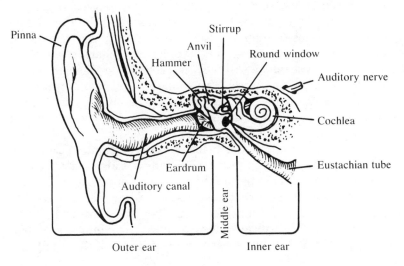

Figure 5.3. The structure of the ear

the eardrum. The flap (technically, the *pinna*) and the *auditory canal* serve to capture sound vibrations in the air and to deliver these vibrations to the *eardrum*. Indeed, one of the major reasons that the ear is most sensitive near 3000 Hz is because the auditory canal has a resonant frequency there.

Sound waves that reach the eardrum cause it to vibrate. The three bones in the middle ear serve to amplify and transmit these vibrations from the eardrum to the inner ear. The names of these bones—the *hammer*, the *anvil*, and the *stirrup*—describe their general shape. One end of the first bone (the hammer) is connected directly to the eardrum, itself, so when the eardrum vibrates, the eardrum drives the hammer. The other end of the hammer is connected to and drives the second bone (the anvil). The third bone (the stirrup) is driven by the anvil and is, in turn, connected to a membrane called the *oval window*. This membrane and another membrane, the *round window,* form the boundary between the middle and inner ears. The middle ear is located in a cavity in the skull, so the delicate transmission system of these three bones is well protected. This cavity is connected to the outside, however, by means of the *Eustachian tube*, which runs from the middle ear to the back of the throat. When atmospheric pressure changes occur, a small amount of air moves into or out of the middle ear thorugh the Eustachian tube. In this way, the pressure on each side of the eardrum is kept the same. You can notice this pressure equalization happening when you change altitude rapidly such as during an airplane flight, a quick elevator trip in a tall building, or a drive in the mountains. At these times, the ears seem to "pop" because the system is designed to work slowly. Thus, the pressure builds up, and hearing is impaired. The sudden pressure relief that occurs when air passes through the Eustachian tube causes the pop. If the system were to work very rapidly, our hearing would be affected because (as we will see later) sound vibrations in the air are, themselves, small pressure changes. On the other hand, if there were no connecting link between the middle ear and the outside, changes in atmospheric pressure would severely distort the eardrum, ruining its ability to vibrate.

The conversion of the mechanical vibrations of sound to nerve signals occurs in the inner ear, which is also well protected within the skull. That part of the inner ear concerned with hearing is contained in the *cochlea*. (The semicircular canals also are contained within the inner ear, but they are used for balance, not hearing.) The spiral-shaped cochlea resembles a sea shell and is filled with fluid. As shown in Figure 5.4, the cochlea is

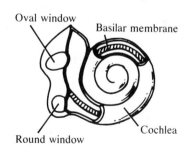

Figure 5.4.   The cochlea and the basilar membrane

divided lengthwise by the *basilar membrane.* The large end of the cochlea is next to the middle ear, where the previously mentioned oval and round windows prevent the fluid in the cochlea from draining into the middle ear. These two windows are located on opposite sides of the basilar membrane. The round window permits movement of the cochlear fluid when changing atmospheric pressure displaces the eardrum and thus displaces the oval window. When, for example, the atmospheric pressure rises, the eardrum and oval window will be depressed. The cochlear fluid then moves through an opening at the end of the *cochlear partition* (basilar membrane) and causes the round window to bulge outward. If there were no round window, then such a change in atmospheric pressure would cause a pressure increase in the cochlear fluid or a stressing of the middle ear bones or both. In any case, the atmospheric pressure change would cause a great deal of pain.

Although the way the ear is protected against atmospheric pressure changes is interesting, our primary purpose in discussing the ear's structure is to see how it detects sound. As mentioned earlier, sound vibrations are passed through the oval window to the fluid in the cochlea. The vibrations in the fluid cause certain regions or places on the basilar membrane to vibrate. The places that vibrate depend upon the frequency and intensity of the sound. Thus, the theory that describes this behavior is called the *place theory* of hearing. Figures 5.5 and 5.6 show differences in the way the basilar membrane vibrates for different sounds. In each of these figures, the cochlea has been unrolled so we more easily can see what is happening. In the different pictures of Figure 5.5, the frequency of the sound is changed while the intensity is kept constant. From this figure, we can see a general tendency that higher frequency sounds vibrate regions of the basilar membrane closer to the oval window. This response of the basilar membrane will be used shortly to help explain the observations we made about masking of sounds. In Figure 5.6 we see the general trend that louder sounds widen the places that are being vibrated on the basilar membrane for a given frequency. By combining ideas in these two sets of diagrams, we arrive at the following conclusion: A high-frequency sound must have a higher intensity than a low frequency sound in order to vibrate the entire basilar membrane.

The ends of the auditory nerves are mounted in the basilar membrane. In particular, in the *organ of Corti,* the auditory nerve ends are attached to hair-like fibers that are moved by the moving fluid and basilar membrane. Thus, when the membrane

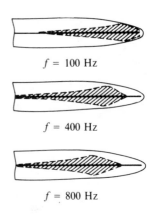

$f = 100$ Hz

$f = 400$ Hz

$f = 800$ Hz

Figure 5.5. The vibration of the basilar membrane at different frequencies (The cochlea is shown uncoiled, and the vibration amplitude is exaggerated for clarity)

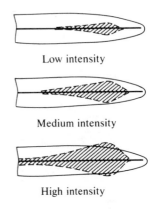

Low intensity

Medium intensity

High intensity

Figure 5.6. The vibration of the basilar membrane at different intensities for the same frequency (The cochlea is shown uncoiled, and the vibration amplitude is exaggerated for clarity.)

vibrates, it causes the nerves to send signals to the brain. Those nerve endings in the places that are vibrating are the nerves that are stimulated. The brain sorts out of these signals, allowing us to identify the pitch, timbre, and loudness of the sounds we hear.

Some aspects of sound detection by the ear are not yet fully understood. As an example of this, we can consider what happens when two pure tones of different frequencies are played together. In addition to hearing the two tones being played, most people will hear several others. These other tones are called *combination tones* or *sum* and *difference tones* because their pitches correspond to frequencies that are simple combinations of the original two frequencies. Table 5 shows how the original two frequencies are combined to produce the tones usually heard. It also includes, as an illustration, the case where the original frequencies are 875 Hz and 1060 Hz. The first additional tone is the difference tone discussed earlier and understood using the concept of beats. The vibrations of the other frequencies, however, are not delivered to the eardrum. Instead, they are created within the ear. Physicists can build mechanical and electrical oscillators that do the same thing. That is, when these oscillators are driven by two frequencies, new frequencies will appear in just these combinations. The lack of understanding occurs when we try to identify those parts of the ear that are behaving like the mechanical and electrical models.

Table 5.   Combinations tones heard when two pure tones are sounded (Original tones for example are f_1 = 875 Hz and f_2 = 1060 Hz.)

Combinations of Original Tones	Example Frequencies
$f_{c1} = f_2 - f_1$	185 Hz
$f_{c2} = 2f_1 - f_2$	690 Hz
$f_{c3} = 3f_1 - 2f_2$	505 Hz

On the other hand, the place theory of hearing explains masking rather well. As noted earlier, a low frequency, high intensity sound tends to vibrate the entire basilar membrane. Thus, the detection of a lower intensity, high frequency sound would be difficult. High intensity, high frequency sounds, however, vibrate only that part of the basilar membrane near the oval window. Thus, a low intensity, low frequency sound can be detected by the vibration it produces at the far end of the basilar membrane.

Summary

We describe sound in either of two ways. The subjective, psychological variables of pitch, loudness, and tonal quality depend primarily on the objective, physical variables of frequency, intensity, and vibration recipe, respectively. However, other relationships also occur. An example that we discussed was how loudness depends on frequency as well as intensity.

The ear is a complicated oscillating system. It is designed to respond to sound wave vibrations. The normal, young ear is sensitive to frequencies between 20 and 20,000 Hz (with a maximum sensitivity around 3,000 Hz), and it responds to intensities between 10^{-12} and 1.0 watts per square meter. As the ear ages or is subjected to loud sounds, it deteriorates, and these ranges decrease. Differential pitch discrimination (the ability to detect frequency changes) approximately follows Weber's law above 1,000 Hz: The ear can detect changes of 0.3 percent in this frequency range. Below 1,000 Hz, however, the ear needs an absolute change of 3 Hz to detect a difference. The ability to detect changes in intensity (differential loudness discrimination) also approximately follows Weber's law. Loud sounds require about a six percent change, while soft sounds require nearly a 100 percent change for detection. A tone of one frequency can mask one of another frequency, and low frequency tones mask high frequency ones more easily than vice versa.

The structure of the ear determines its performance in detecting sound. Pressure vibrations moving through the air pass down the auditory canal and strike the eardrum, causing it to vibrate. These vibrations are transmitted through three bones (the hammer, anvil, and stirrup) in the middle ear to the oval window and then into the cochlea. The vibrating fluid within the cochlea drives the nerve endings mounted in the organ of Corti on the basilar membrane, and nerve signals are sent to the brain. According to the place theory of hearing, the locations of the basilar membrane that vibrate determine the pitch that is perceived, and the amplitude and width of the vibrating areas determine the loudness. Since high frequency sounds do not cause the tip of the basilar membrane to vibrate, we can understand why it is difficult for high frequency sounds to mask low frequency ones.

Suggested Readings

Both the books *Waves and the Ear*, by Willem A. Van Bergeijk, John R. Pierce, and Edward E. David, Jr. (Garden City, N.Y.: Anchor Books/ Doubleday, 1960) and *The Speech Chain*, by Peter B. Denes and Elliot N. Pinson (Bell Telephone Laboratories, 1963) discuss the process of hearing. Each book goes into greater detail than this book, but the level of presentation is about the same. *The Speech Chain* also describes the production of sound by the human voice, which we have not covered at all.

Most books on the physics of music contain a section on the workings of the human ear. This section in *Musical Acoustics,* by Donald E. Hall (Belmont, Calif.: Wadsworth Publishing Company, 1980), is very informative.

Review Questions

Five pure tones (which we will call A, B, C, D, and E) are sounded with the physical characteristics shown in Table 6. Use Table 6 to answer review questions 1–4.

Table 6.   For use with review questions 1–4.

Tone	Frequency (Hz)	Intensity (W/m^2)
A	1000	10^{-10} (0.0000000001)
B	100	10^{-10} (0.0000000001)
C	100	10^{-7} (0.0000001)
D	10,000	10^{-8} (0.00000001)
E	15,000	10^{-8} (0.00000001)

1. For a normal young ear, which tone is probably the loudest?

    a. A

    b. B

    c. C

    d. D

    e. E

2. For a sixty-five-year-old person, which tone or tones are probably inaudible?

    a. B

b. E

c. B and E

d. A, B, and E

e. A, B, D, and E

3. For a normal young ear, which tone or tones sound about equal in loudness to tone A?

a. B

b. C

c. D

d. E

e. C, D, and E

4. If tones A and C are played together and their intensities are increased proportionately, what will happen?

a. Tone A eventually will mask tone C.

b. Tone C eventually will mask tone A.

c. We will continue to hear both tones A and C.

5. A complex vibration has many harmonics. Specifically, it consists of the frequencies 200 Hz, 400 Hz, 600 Hz, 800 Hz, 1000 Hz, 1200 Hz, etc. What will happen to the pitch of this tone when all frequencies below 500 Hz are electronically removed?

a. It will change from a pitch associated with 200 Hz to one corresponding to 400 Hz.

b. It will change from a pitch associated with 200 Hz to one corresponding to 500 Hz.

c. It will change from a pitch associated with 200 Hz to one corresponding to 600 Hz.

d. It will not change; the pitch will correspond to 200 Hz in both cases.

6. The same experiment as that in question 5 is repeated with a vibration consisting of only odd harmonics, specifically 200 Hz, 600 Hz, 1000 Hz, 1400 Hz, etc. What happens to the pitch in this case when all frequencies below 500 Hz are removed?

    a. It will change from a pitch associated with 200 Hz to one corresponding to 400 Hz.

    b. It will change from a pitch associated with 200 Hz to one corresponding to 500 Hz.

    c. It will change from a pitch associated with 200 Hz to one corresponding to 600 Hz.

    d. It will not change; the pitch will correspond to 200 Hz in both cases.

7. Two tones are sounded alternately. One has a frequency of 2000 Hz, and the other has a variable frequency. The frequency of the second is changed until we can just tell that the two are different. Then, the two tones are sounded together. How many beats per second will be heard?

    a. 0.3

    b. 3

    c. 6

    d. 30

    e. 60

8. Mechanical vibrations are transformed to nerve impulses in the

    a. middle ear.

    b. Eustachian tube.

    c. auditory canal.

    d. semicircular canals.

    e. cochlea.

9. The correct order in which mechanical vibrations pass through the parts of the ear is

    a. eardrum, cochlea, hammer, stirrup, and oval window.

    b. eardrum, hammer, anvil, stirrup, and cochlea.

    c. eardrum, Eustachian tube, oval window, and round window.

    d. eardrum, anvil, Eustachian tube, and organ of Corti.

    e. eardrum, cochlea, basilar membrane, and middle ear.

6

Introduction to Waves

Important Concepts

- For all types of waves,

    1. the wave speed is determined by the medium.

    2. the wave speed, frequency, and wavelength always satisfy the relation $v = f\lambda$.

- The medium moves in a direction perpendicular to the wave motion for a transverse wave, parallel to the wave motion for a longitudinal wave, and along circular or elliptical paths for water waves.

- Transverse and water waves are made up of a series of crests and troughs; condensations and rarefactions are the analogous qualities of a longitudinal wave.

- When the water depth is greater than one-half the wavelength $(D > 0.5\lambda)$, we have deep water waves with $v \propto \sqrt{\lambda}$.

- When the water depth is less than one-twentieth the wavelength $(D < 0.05\lambda)$, we have very shallow water waves with $v \propto \sqrt{D}$.

- Sound waves are longitudinal and travel at high speeds in air, at higher speeds in lighter gases such as helium and hydrogen, and at even higher speeds in liquids and solids.

- Since water waves do not form sine curve shapes, we do not use the concept of amplitude with them. Rather, the wave height is defined as the vertical distance from the top of a crest to the bottom of a trough.

- Waves on ideal strings and suspended slinkies, sound waves, and very shallow water waves are non-dispersive. That is, the wave speed for them is the same for waves of all frequencies (or wavelengths).

- Dispersion occurs for deep-water waves (where long wavelength waves travel faster) and for waves on stiff strings (where long wavelength waves travel slower).

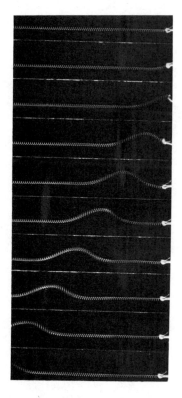

A pulse moving along a suspended slinky. The same motion occurs for a pulse on a stretched string. (PSSC Physics, 2nd edition, 1965; D.C. Heath and Company with Education Development Center, Inc. Newton, MA.)

Now, we are ready to start our investigation of waves. For most of us, the word "waves" brings to mind a particular image. I am speaking, of course, about waves on bodies of water such as the ocean, ponds, or lakes. We term these waves *traveling waves* because we can watch them moving from one location to another. Before we study these water waves, however, we will look first at waves that travel along a string. There are two reasons for looking at waves on strings first: Waves that travel along strings are conceptually simpler, and we can bring them into the laboratory more easily. We also will investigate other types of waves such as waves on suspended slinkies and sound waves in this chapter. We will focus mainly on the similarities and differences among these various types of waves.

One of the differences among these waves is that the waves travel in different dimensions. The waves on a string or a suspended slinky are *one-dimensional*—or, we could say they travel in a one-dimensional space—because we need only one number to locate each point along the string. (This number could, for example, be the distance along the string from one end.) Water waves travel in a *two-dimensional* space because we need two numbers to locate each point on the surface of the water. (For example, one number tells how far north the point is from an origin, and the other number tells how far east the point is from the origin. Finally, sound and light waves are *three-dimensional* because we need three numbers to describe the location of a point in the space where they travel. The third number tells how high the point is.

It becomes progressively more difficult for us to describe wave motion in each of these cases. For example, waves traveling on a string can move only to the left or to the right. But two-dimensional or three-dimensional waves can move in many directions at the same time. We referred to the conceptual simplicity of waves on a string before, and, indeed, it is a good reason for starting our introduction to waves with them.

Almost all types of waves are similar in that they are vibrations being transported through a medium, as described in Chapter 1. I remind you that the medium is the material through which the wave travels. Thus, for a wave on a string, the medium is the string itself. Sound waves can travel through various media such as air, water, or solids. Light and other electromagnetic waves are exceptional in that they can travel through a vacuum. They also can travel through various media such as air, water, and glass.

Traveling Waves on a String

We will start our discussion about waves on a string by considering a very long string. We start with a long string so that we do not have to worry about what happens when a wave reaches the end. (The end is so far away from the wave that as far as we are concerned, the wave never gets there!) We can produce waves on this string by jiggling one end of it. In order to make the resulting waves simple, we will move the end of the string by attaching it to a simple harmonic oscillator. As a result, the wave on the string will be shaped like a sine curve at every instant of time. As the wave travels along the string, each point on the string moves up and down as the wave passes. The string never moves in the direction of the wave. The successive pictures in Figure 6.1 show the production and movement of a wave on a long string. The string is attached to a mass suspended on a spring, and when the mass moves, it pulls the end of the string along with it. In picture A, the mass is at its equilibrium position. Since the mass is just starting to move, there is no wave on the string. The time elapsed between each pair of adjacent pictures is one-fourth of the period of the oscillating mass. Thus, in picture B, the mass has moved to the top of its motion, and in C, the mass has moved back to its equilibrium position, etc.

We clearly can see from this set of pictures that the traveling wave consists of a set of *crests* and *troughs*, one following

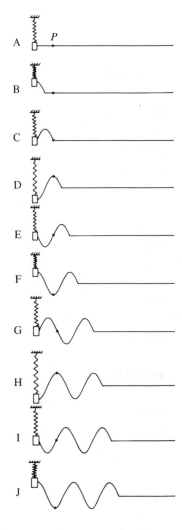

Figure 6.1. A transverse wave traveling along a string

after the other as the wave moves along the string. Figure 6.1 also keeps track of a particular point on the string (labeled P in picture A), so we can follow its motion as the wave moves by. This point (as with every other point on the string) moves up and down executing periodic motion as the wave passes by. Since the original disturbance producing the wave executed simple harmonic motion, each point on the string will do so also. Thus, we can talk about the frequency and period of the string's motion while the wave moves by. We also will refer to these factors as the *frequency* and *period* of the wave, although it is actually each point on the string that is moving with periodic motion. In the same way, the amplitude of each point's vibration is termed the *amplitude* of the wave.

While each part of the string is moving with simple harmonic motion, new concepts here require additional vocabulary. First, we use the concept of *speed* (or *velocity*) to describe how fast the wave is moving. A speed always is quoted as a distance divided by a unit of time. Our most common experience with speed occurs with the automobile. Here, the unit of time is the hour, and the unit of distance is usually the mile, but sometimes the kilometer. Speed tells us how far something moves during that unit of time, provided the speed does not change. Thus, if you drive your car for one hour at a constant speed of 30 miles per hour, you will have traveled 30 miles. The same thing is true for waves. A wave traveling at a constant speed of 5 meters per second (also written 5 m/s) will travel 5 meters during each second. These concepts indicate a way to determine the speed of a moving object (or wave). By measuring the time required for the object to move a known distance (or by measuring the distance traveled in a known time), we can calculate the object's speed. The speed is the distance traveled divided by the time. Using symbols, we can write the following equation to express this calculation:

$$v = \frac{D}{t} \qquad \textbf{(6.1)}$$

In this equation, D stands for the distance traveled, t for the time, and v for the speed. (The symbol v is derived from the word *velocity*.)

The speed of every wave is determined by the medium (or material) through which the wave is moving. Thus, the speed of the wave on the string is determined by the string itself. This is a very important concept; we will use it again and again as we look at other types of waves. To emphasize, I shall repeat: The speed

of any wave is determined by the medium in which it is traveling. It follows that if the medium changes, then the wave speed
will change.

The second concept about waves I wish to describe is
expressed with the term *wavelength*. The wavelength of a wave
on a string is the distance from a point on one cycle of the wave to
the analogous point on an adjacent cycle. One of the easiest
points to use for this determination is a crest. Thus, the wavelength would be the distance from one crest to the next. However, I wish to stress that the choice of the crest as a point to
measure from is purely arbitrary; any other point could be used.
Two different choices for measuring the wavelength are shown
in Figure 6.2. The symbol λ (a Greek letter, named lambda) is
widely used to stand for the wavelength of waves, so we will use
the symbol also.

There is a fundamental relationship between the frequency,
wavelength, and speed of a wave that occurs because of the way
these terms are defined. Let's look again at Figure 6.1, particularly at point P on the string. Note that as one complete wavelength of the wave passes by this point (which occurs between
pictures C and G, for example), the wave executes one cycle of its
simple harmonic motion. Thus, the wave has moved a distance
equal to one wavelength during a time that is equal to one period. By using the definition, we see that the speed of the wave is
equal to the wavelength divided by the period. Further, since the
period is the reciprocal of the frequency, we find the speed of the
wave is the product of the wavelength and the frequency. These
relationships are shown using symbols in Equation 6.2.

$$v = \frac{\lambda}{T} = f\lambda \qquad \textbf{(6.2)}$$

This is a very fundamental relationship and is always true. The
frequency of the wave, you will remember, is determined by the
source of the vibration. The speed of the wave is determined by
the medium through which the wave moves. Thus, the wavelength adjusts itself so that Equation 6.2 is always satisfied.
This behavior happens for all types of traveling waves.

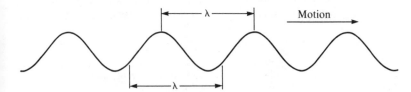

Figure 6.2. The wavelength
of a traveling, transverse wave

As I have pointed out, the speed of a wave depends upon the medium through which it moves. For an ideal string, two things determine the speed of a wave on it: the weight* of the string and how tightly the string is stretched (that is, its tension). For a heavier and thicker string, the wave moves more slowly, and increasing the tension causes a higher wave speed. Using symbols, we can write this relationship concisely as

$$v = \sqrt{\frac{T}{d}} \qquad \text{(6.3)}$$

where v is the wave speed, T is the string tension, and d is the linear density of the string. This expression is presented only so we see clearly that the wave speed depends on the square root of the tension and, inversely, on the square root of the "weight." Thus, if we want to double the wave speed, we must quadruple its tension.

It is possible to change the string's thickness along its length. For example, fly-fishing lines are tapered to improve their performance in casting. For such a string, a wave will change its wavelength as the wave moves along the string. Note that the frequency of vibration remains the same; the speed decreases as the string gets thicker, causing a shorter wavelength. It is also possible to vary the tension in a string by hanging it vertically from one end. The tension is large at the top of such a string and is zero at the bottom end. Thus, waves traveling down the string will slow down, causing a decrease in the wavelength. This decrease in the wavelength squeezes the energy carried by the wave into a smaller space, so the amplitude of the wave increases as the wavelength gets smaller. For most strings, however, both the tension and the weight stay the same, so the wave speed along these strings is constant.

For such an ideal string, as we have been discussing, the wave speed is independent of the frequency or wavelength of the wave. A violin string is flexible enough that it is nearly an ideal string. A piano string or a guitar string is made of wire, so it is not this flexible. The stiffness of the "string," in this case, causes low frequency (long wavelength) waves to travel more slowly than high frequency ones. We will see in Chapter 8 that this condition explains why a piano string's normal modes are not harmonics.

---

*As mentioned on the footnote on page 71, the correct variable is the linear mass density or mass per unit length. We usually can think of the string's weight, however, because it is directly related to the density.

Transverse and Longitudinal Waves

The wave traveling along a string is called a *transverse wave* because as the wave moves along the string, the string itself, moves from one side to the other. In contrast to this type of wave, we have *longitudinal waves*, where the material moves backward and forward in the same direction in which the wave is traveling. Thus, in a transverse wave, the medium moves in a direction perpendicular (or transverse) to the wave motion as the wave passes by. For a longitudinal wave, the medium moves parallel to (or along) the direction of wave motion. Longitudinal waves do not have crests and troughs like those of transverse waves. Instead, the medium gets squeezed together or stretched apart as the wave moves by. We call a point where the medium is squeezed together a *compression* or *condensation*. Points where the material is stretched apart are called *rarefactions*. Thus, a traveling longitudinal wave is a sequence of condensations and rarefactions following along, one after the other, just as a transverse wave is a series of crests and troughs moving along, one after the other.

The speed of a longitudinal wave is defined in the same way as the speed of a transverse wave: It is the distance traveled during a certain amount of time divided by that time. The wavelength of a longitudinal wave is determined in a way that is analogous to the method of determining the wavelength for a transverse wave. That is, the wavelength is the distance between similar points on adjacent cycles of the wave. For example, the distance between two adjacent condensations in a longitudinal wave is the wavelength of that wave. The frequency or period of a longitudinal wave also is found in the same way as for a transverse wave: The period is the amount of time it takes for the material to go through a complete cycle as the wave passes. Finally, it is also true that the speed of the wave is equal to the frequency times the wavelength. Thus, the only difference between longitudinal and transverse waves is the direction in which the medium moves.

If a slinky* is suspended by strings in such a way that its coils are stretched apart, then we can send a traveling longitudinal wave along the slinky by alternately pulling and pushing on one end of the slinky. Figure 6.3 shows a longitudinal wave on a stretched slinky; the wave is produced by a horizontally mov-

---

*A slinky is a large diameter spring made with small diameter wire. It originally was sold as a toy, but now finds great practical use in physics education.

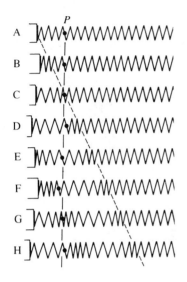

Figure 6.3. The movement of a longitudinal wave along a suspended slinky

ing mass. In picture A, the mass is at its equilibrium position (but just starting to move), and there is no wave on the slinky. The time that elapses between each pair of adjacent pictures is one-fourth of the period of the oscillating object attached to the end of the slinky. Thus, in picture B, the object has moved to the right end of its motion, and in C, it's back to its equilibrium position, etc.

The vertical line allows us to see more easily the motion of the point P, which is attached to the fourth coil of the slinky. As the longitudinal wave passes, the point P moves backward to meet the approaching compression, travels forward with the compression through the equilibrium point, and then moves backward again through the next rarefaction. You can follow this sequence of events most easily in pictures C–G. The diagonal, dashed line follows a single compression as it moves along the suspended slinky.

In addition to a slinky, a longitudinal wave also could be sent along a string (either ideal or real). However, the wave would be difficult to see because the string does not stretch very much, so the displacement of each point on the string as the longitudinal wave went by would be very small. The wave also would travel very rapidly. A longitudinal wave on a suspended slinky travels slowly enough so that we can use a stopwatch to measure the time it takes for the wave to move a distance of several meters. The speed of such a wave is typically about one meter per second. We also could send a transverse wave along the slinky and measure its speed in the same way. Comparing the speeds of the longitudinal and transverse waves on the same slinky would show the longitudinal wave is faster. For many types of media, longitudinal waves are faster than transverse waves. Other examples of this include waves on strings and waves moving through the earth (such as those produced by an earthquake).

Sound as a Longitudinal Wave

Experiments show that sound waves are also longitudinal waves. As a sound wave moves through air, the molecules of the air are alternately compressed and spread apart. This displacement of the molecules produces a local change in the pressure of the air. We can observe the movement of sound waves through air by the displacement of the molecules or by the change in pressure as the wave passes. Sound waves also can travel through liquid and solid media. In all cases, sound is a longitudinal wave, so the particles of the medium vibrate back and forth parallel to the direction of the wave's motion.

Sound waves moving through the air travel considerably faster than waves on a slinky. Thus, we cannot use a stopwatch to measure the time it takes for sound to move a measured distance. We can, however, measure the speed of a sound wave in air using two microphones hooked to an oscilloscope. We arrange this equipment so that a dot starts across the oscilloscope screen when the sound reaches one microphone. When the sound reaches the second microphone, the dot will register a displacement on the oscilloscope screen. Since we know the speed of the dot moving across the screen, we can measure the time interval required for the sound wave to travel between the two microphones. The physical arrangement of this equipment is shown in Figure 6.4 and Figure 6.5 shows a typical trace on an oscilloscope screen for this experiment. In the trace of Figure 6.5, the speed of sound through air was 333 meters per second. (The sound traveled 2 meters in 0.006 sec.) The value usually quoted for the speed of sound in air is 340 meters per second (or 1100 ft per second). The value varies according to the condition of the air; for example, sound waves travel faster through warmer air. An even greater difference in the speed of sound occurs when the material is changed. Replacing the air with a lighter gas such as helium or hydrogen will give a speed of sound of about 1000 to 1500 meters per second, which is three to five times faster than in air. The speed of sound in a solid material (a material that is much stiffer than a gas) ranges upward from about 5000 meters per second. In all cases, however, the speed of sound (or, more technically, the speed of the sound waves) is determined by the medium through which the waves are traveling.

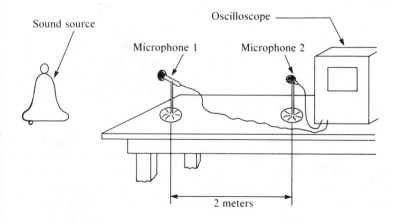

Figure 6.4. Two microphones and an oscilloscope set up to measure the speed of sound

Figure 6.5. The oscilloscope trace obtained in the experiment of Figure 6.4.

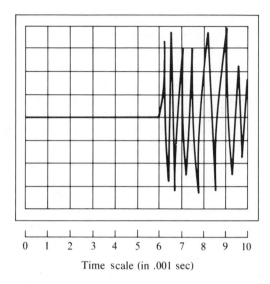

Time scale (in .001 sec)

Even though sound travels rapidly through the air, light moves much, much faster. Over a distance of several miles, the movement of light is practically instantaneous. Since the speed of sound is known, it is possible for you to estimate how far you are from a lightning flash during a thunderstorm. The flash of lightning and the thunder are created at the same time. The light immediately arrives at your eye, so the time interval between when you see the flash and hear the thunder is the time required for the sound to reach you. As noted before, sound travels about 1100 feet per second, which is about ⅕ mile per second. Thus, counting seconds during the time interval and multiplying by ⅕ will give the distance to the lightning in miles.

Water Waves

A third type of traveling wave occurs on the surface of water. We will call these water waves, but they are more correctly called gravity waves because gravity provides the force needed to make the waves move across the surface. (Another type of water wave produced by surface tension and called, therefore, a surface tension wave also occurs. However, this type of wave has a very short wavelength and is not of interest to our discussion.) Water waves are neither transverse nor longitudinal; the particles of water move in circles as the wave passes by. Since the water does not execute simple harmonic motion, we do not use the term amplitude to describe the water's vertical displacement. Instead, we define the wave height, *H,* to be the vertical distance from the bottom of a trough to the top of a crest. (See Figure 6.6.) Thus, the wave height of a water wave is similar to twice the amplitude of a transverse wave, not to the amplitude itself.

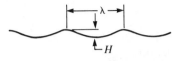

Figure 6.6.   A water wave

When we look at waves on lakes or oceans, it is often difficult to see any regular motion at all. Only after tanks were built in laboratories for the observation of controlled waves were scientists able to understand the behavior of water waves. We now know that the water under the surface moves as a wave passes, but only to a depth about equal to one-half the wavelength. Because of this partial movement, the bottom of the ocean (or lake, pond, etc.) cannot affect the wave if the water is deeper than one-half the wavelength. Consequently, we divide water waves into three categories, depending upon the depth of the water as compared to the wavelength. When the water is deeper than one-half (½ or 0.5) the wavelength, we term the waves *deep water waves*. When the depth of the water is less than one-twentieth (¹⁄₂₀ or 0.05) of the wavelength, we term the waves *very shallow water waves*. Waves falling in the range between these two extremes are termed *shallow water waves*. We divide water waves into these three categories because the wave speed behaves differently in each case. Scientists have developed a complicated formula* for calculating the speed of water waves; the formula depends on the depth of the water and the wavelength of the wave. This complicated formula is shown in Equation 6.4, and it must be used to calculate the speed of waves that fall into the intermediate (shallow-water-wave) category.

$$v = \sqrt{\frac{g\lambda}{2\pi} \tanh\left(\frac{2\pi D}{\lambda}\right)} \qquad \textbf{(6.4)}$$

The formula also works for the other two types of water waves, but for these cases, the formula can be simplified, and we will use the simplified relations. We will concentrate our discussion on waves in these two categories.

When the depth of water is greater than one-half the wavelength (deep water waves), we say the wave does not feel bottom. Thus, if a water wave has a wavelength of 6 meters (about 20 feet) and the depth of water is at least 3 meters, the wave does

A ripple tank, which can be used to observe the movement of water waves. The bar suspended from springs at the far end of the tank is used to generate plane water waves.

*The detailed formula is shown for completeness only; we will not be using it. For those interested students, here are the meanings of all the symbols: g = gravitational acceleration, π = ratio of circumference to diameter of a circle, D = water depth, and λ = wavelength. Finally, tanh is the hyperbolic tangent function, which is similar to the sine or cosine function.

not "know" where the bottom is. There is no difference as far as the wave is concerned between a water depth of 3 meters and a water depth of 300 meters. The speed of the wave is determined only by its wavelength: Longer wavelength waves travel faster than shorter ones.

It is still true that the wave speed is determined by the medium. Only, in this type of medium (that is, deep water), the wave speed varies for different wavelength waves. In this case, the formula for the wave speed given in Equation 6.4 simplifies to

$$v = \sqrt{\frac{g}{2\pi}\lambda}$$ **(6.5)**

and we see that the wave speed is proportional to the square root of the wavelength. Evaluating the proportionality constant allows us to be more specific; we find $v = 1.25\sqrt{\lambda}$, with λ measured in meters and v in meters per second. While we can use this formula to calculate wave speed, the important point is that wave speed varies with the square root of the wavelength.

This is the same type of characteristic that we mentioned briefly in connection with waves on stiff strings. Whenever the speed of a wave depends upon its wavelength (or, therefore, its frequency), we have *dispersion*. When waves travel through a dispersive medium (one that causes dispersion), waves of different wavelengths will separate from each other. This separation happens with waves on stiff strings and with deep water waves, and we will find it again when we study light. We will discuss dispersion in greater detail in Chapter 11.

When water waves travel across the surface, they do not have the shape of a sine curve like the transverse waves on a string that we were looking at. Instead, the crests of water waves are sharper, and the troughs are broader. As mentioned earlier, when a wave moves by a point, particles on the surface of the water at that point move in a circle. For deep water waves, particles of water under the surface also move on circular paths. The diameters of these circular paths decrease as we move farther under the water. Finally, at a depth equal to about half the wavelength of the wave, the water does not move at all. This behavior is shown in Figure 6.7, which also shows the direction in which the particles of water move around the circle. As can be seen in the picture, the water moves in the same direction as the wave when a crest is passing and moves backward relative to the wave motion when a trough passes by. But during one complete cycle, there is no net movement of the water.

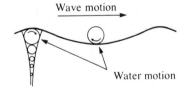

Figure 6.7. Water movement during passage of deep water waves

Very shallow water waves are different from deep water waves in several ways. First of all, very shallow water waves "feel the bottom." As a matter of fact, they appear to drag on the bottom because as the water gets shallower, the waves slow down. The water depth is the only thing that determines the speed of a very shallow water wave. In this case, the general equation (6.4) simplifies to

$$v = \sqrt{gD} \tag{6.6}$$

Again, we can evaluate the constant, and we have $v = 3.1\sqrt{D}$, with the water depth, D, measured in meters and v in meters per second. Note that the wave speed does not depend upon the wavelength of the wave, so there is no dispersion for very shallow water waves. There is also a difference in the way the water moves when shallow water waves pass by. The bottom prevents the water particles beneath the surface from moving along the circular paths these particles would follow if we had deep water waves. The presence of the bottom squeezes the circular path at each depth into an ellipse. On the surface, the paths are almost circular. As we look deeper under the water, the elliptical paths become flatter and flatter until finally on the bottom itself, the path is a horizontal line. At all depths, the length of the water's horizontal movement is the same. This pattern is shown in Figure 6.8, where, again, the direction of movement around the ellipse also is shown. As with deep water waves, the water moves forward as a crest passes and backward as a trough passes by. Although there is no dispersion for very shallow water waves, there is another effect that is important. The waves can change both speed and direction as the depth of the water changes.

In order to describe these effects, we first must recognize the two-dimensional character of water waves and must define some new terminology. Water waves travel across the surface of water, so they have width or breadth as well as length. For example, if we drop a stone into a pond of still water, circular waves will spread out from the point where the stone hit. (See Figure 6.9.) We recognize the distance between any pair of adjacent crests as the wavelength, but here, the wave extends completely around the circle. Also, the wave is not moving in only one direction; it is moving in all directions at once. In order to more fully describe this wave, we define the term *wavefront*. The easiest way to construct a wavefront is to draw a line along the top of a single crest of the wave. Each crest produces a different wavefront, of course. Figure 6.10 shows the wavefronts constructed in this way for the waves in Figure 6.9. These wave-

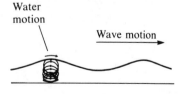

Figure 6.8. Water movement during passage of shallow water waves

Figure 6.9. Circular waves produced by dropping a stone into still water (Fundamental Photographs, New York)

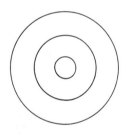

Figure 6.10. Wavefronts for the waves shown in Figure 6.9

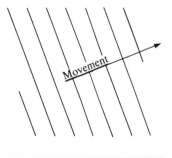

Figure 6.11. Wind-generated waves often have straight wavefronts

fronts are circular, but those generated by wind blowing across the surface of a lake or ocean are usually straight lines such as those shown in Figure 6.11. Other shapes are, of course, possible; the important point is that by including information about the wavefront shape, we are able to describe additional aspects of the wave. It is also helpful to recognize that a wave always moves in a direction perpendicular to its wavefront. Thus, the waves in Figure 6.9 spread in all directions at the same time, while those in Figure 6.11 move in only one direction.

We now consider the movement of waves as they approach the shore across a region of water where the bottom slopes gently up to dry land. In water deeper than one-half the wavelength, the waves do not feel the bottom, and their speed is determined by the wavelength. As the water becomes shallower, the waves will start to drag on the bottom and slow down. The waves move slower and slower as the water depth gets to be less and less. For waves heading straight into shore (so their wavefronts are parallel to the shore), the only effects produced by the slower speed are a shorter wavelength and a greater height. Eventually, the waves break, but that subject will be considered in the next chapter. The shorter wavelength occurs because the wave does not travel as far during one period, and the period (and frequency) do not change. This effect is shown in Figure 6.12 through both the cross section in picture A and the use of wavefronts as viewed from overhead in picture B.

In this photograph of water waves, the locations of the wavefronts (along the wave crests) are easily identified. (Photograph courtesy of Steve Lissau.)

When the waves approach the shore at an angle (rather than head-on), they will turn as they slow down. We can see why this happens by referring to Figure 6.13, which is similar to Figure 6.12B, except the waves are approaching the shore at an angle in Figure 6.13. The end of each wavefront that is closest to the shore will be in shallower water and, therefore, will travel slower than the end in deeper water. Thus, the end near shore does not move as far as the deeper end during the same amount of time. A useful metaphor for this situation is to think of each wave crest as a row of marching soldiers. The row of soldiers is lined up shoulder-to-shoulder along the wavefront. The deep water contains tall soldiers taking long strides as they march

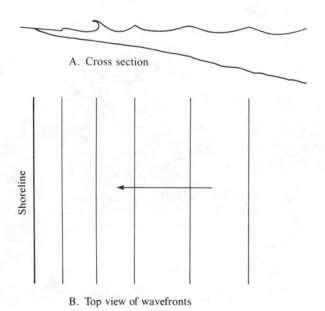

A. Cross section

Shoreline

B. Top view of wavefronts

Figure 6.12. Waves moving head-on into a beach over a gently sloping bottom

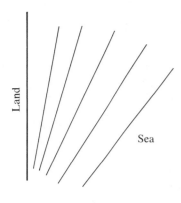

Figure 6.13. The refraction of water waves approaching the shore

along, while the shallow water hoids short soldiers taking smaller steps. This setup causes the entire row of soldiers to revolve around the slower end. For the waves, this setup means the wavefronts become more nearly parallel to the shore as they move toward land, and the waves end up moving straighter into the shore. This change in wave direction, caused by its change in speed, is termed *refraction*. Refraction will be discussed more fully in Chapter 11. More details of water waves will be covered in the next chapter.

Summary

Traveling waves are the movement of vibrations from one location to another. We looked at three types of waves that are classified according to how the medium moves as a wave passes: transverse, longitudinal, and water waves. Examples of transverse waves that we discussed were waves on a string and on a suspended slinky. Each of these examples are one-dimensional because a location in the medium can be specified using only one number. Water waves (or, more properly, gravity waves) move in the two-dimensional medium of the water's surface. Sound is a longitudinal wave moving in the three-dimensional space in which we live. Transverse and water waves contain a series of crests and troughs, while longitudinal waves have compressions and rarefactions.

For all types of waves, the wavelength is the distance between analogous points in adjacent cycles of the wave. Also,

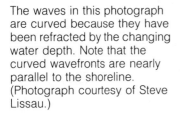

The waves in this photograph are curved because they have been refracted by the changing water depth. Note that the curved wavefronts are nearly parallel to the shoreline. (Photograph courtesy of Steve Lissau.)

the product of the frequency and the wavelength of a wave is always equal to the speed of the wave. The wave's frequency is determined by the source, and the wave's speed is determined by the medium through which it moves. Thus, the wavelength must adjust itself to satisfy its relation with the frequency and speed.

The speed of a wave on a string is determined by the string's tension and mass per unit length (roughly the weight): The wave goes faster with tighter or lighter strings. Sound waves move faster through warmer air and even faster through less dense gases such as helium and hydrogen or through solid materials. A complicated expression was given that shows the dependence of water wave speed on various factors. Fortunately, this expression simplifies for two extreme cases: When the water depth is greater than one-half the wavelength, we have deep water waves, and the wave speed depends only on the wavelength of the wave. The other extreme case is when the depth is less than one-twentieth of the wavelength, so we have very shallow water waves, and the wave speed depends only on the depth.

Suggested Readings

Waves in general are discussed in the first few chapters of *Sound Waves and Light Waves*, by Winston E. Kock (Garden City, N.Y.: Anchor Books/Doubleday, 1965). Applications are obviously in the areas of sound and light.

Water waves are described more extensively in the book *Waves and Beaches*, by Willard Bascom (Garden City, N.Y.: Anchor Press/Doubleday, 1980). The level of sophistication in *Waves and Beaches* is about the same as that used in this book, but the author's first-hand experiences with ocean wave study give *Waves and Beaches* a wonderful flavor.

Review Questions

Questions 1–3 are based on this statement: A transverse wave with a frequency of 400 Hz travels along a string at a constant speed of 100 m/s. The string length is 20 meters.

1. How long does it take for this wave to move from one end of the string to the other?

   a. $\frac{1}{100}$ sec

   b. $\frac{1}{20}$ sec

   c. $\frac{1}{5}$ sec

   d. 5 sec

   e. 20 sec

2. What is the wavelength of the wave?

   a. $\frac{1}{400}$ meter

   b. $\frac{1}{5}$ meter

   c. $\frac{1}{4}$ meter

   d. 4 meters

   e. 5 meters

   f. 20 meters

3. What is the period of this wave?

   a. $\frac{1}{400}$ sec

   b. $\frac{1}{100}$ sec

   c. $\frac{1}{4}$ sec

   d. 4 sec

   e. 20 sec

4. A longitudinal wave with the same frequency as the transverse wave discussed in questions 1–3 is sent along the same string as described in those questions. Compared to the transverse wave,

a. the speed of the longitudinal wave is greater, and its wavelength is longer.

b. the speed of the longitudinal wave is greater, and its wavelength is shorter.

c. the speed of the longitudinal wave is less, and its wavelength is longer.

d. the speed of the longitudinal wave is less, and its wavelength is shorter.

e. the longitudinal wave has the same speed and wavelength because the speed depends on the medium.

5. You are watching a carpenter nail a board on a house 165 meters away. How long is the time interval between when you see him hit the nail and when you hear the sound of his hitting?

a. Zero (The sound arrives at the same time that you see him hit the nail.)

b. ½ sec

c. 1 sec

d. 2 sec

e. 10 sec

6. Three seconds after you see a lightning flash, you hear the thunder. How far away was the lightning?

a. ⅕ mile

b. ⅗ mile

c. 1 mile

d. 3 miles

e. 5 miles

7. A water wave has a height of 2 meters and a wavelength of 40 meters. What is the minimum water depth this wave must have to be considered a deep water wave?

a. 1 meter

b. 2 meters

c. 4 meters

d. 20 meters

e. 40 meters

8. A water wave with a 5-meter wavelength and a ½-meter height moves from an area where the water depth is 6 meters to one where the depth is 3 meters. The wave will

a. decrease in speed and increase in height.

b. decrease in speed and decrease in height.

c. increase in speed and increase in height.

d. increase in speed and decrease in height.

e. not change.

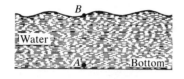

Figure 6.14. For use with review question 9

9. Figure 6.14 (which is not to scale) shows a wave moving across the surface of water 2 meters deep. The wavelength is 50 meters; therefore, the motion of the water at points A and B is, respectively,

a. circular and elliptical.

b. horizontally back and forth, and circular.

c. elliptical and elliptical.

d. impossible to determine because the wave direction is not shown.

10. A long wavelength deep water wave has a speed of 20 m/s. What is the wavelength of this wave?

a. 1.5 meters

b. 3.3 meters

c. 55 meters

d. 124 meters

e. 256 meters

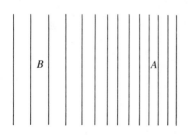

Figure 6.15. For use with review question 11

11. Figure 6.15 shows a series of wavefronts for water waves. Which one of the following statements is most correct?

a. The water is shallower at A than at B, and the waves are moving from A to B.

b. The water is shallower at A than at B, and it is impossible to tell which way the waves are moving.

c. The water is shallower at A than at B, and the waves are moving from B to A.

d. The water is shallower at B than at A, and the waves are moving from A to B.

e. The water is shallower at B than at A, and it is impossible to tell which way the waves are moving.

f. The water is shallower at B than at A, and the waves are moving from B to A.

7

Ocean Waves

Important Concepts

- Water waves are formed by wind, underwater disturbances, and the gravitational pull of the moon and sun.

- The size of wind-generated waves depends primarily on the wind speed. The size also depends on the distance over which the wind blows (the fetch) and the duration of the wind.

- Water waves break when they become too steep. In deep water, the waves break when the steepness (wave height divided by wavelength) gets up to $1/7$. Waves moving into shore break when their height is $3/4$ the water depth.

- Water waves satisfy the principle of superposition only when they are not breaking.

- Surf beat is an example of beats, which are produced by two sets of waves with different frequencies superposing.

- Mass transport occurs only with breaking waves.

Waves occur on the surface of many bodies of water such as ponds, lakes, streams, rivers, and oceans. Because of its size, there is greater variety in the waves on an ocean than on other bodies of water. Thus, the discussion of ocean waves in this chapter encompasses all of the types of water waves we wish to discuss, and the ideas can be transferred easily to other situations.

In our study of ocean waves, we will look at both their birth and death. We start by looking at the three different physical processes that actually produce the wide variety of waves we observe on the ocean. Then we will look at the way waves expire when they break—either far at sea or against a coast. We also will look at some special topics of ocean behavior that waves help us to understand.

Sources of Ocean Waves

We can classify ocean waves into three different categories, depending upon the waves' source. The most common type of wave is generated by the wind blowing across the water's surface. These wind-generated waves come in a wide range of sizes, from little ripples to great storm waves. A second way of creating waves is through some sort of underwater disturbance, such as an earthquake or a landslide. A wave generated in this way is termed a *tsunami*. The third category of wave is the *tide*, which is created by the gravitational pull of the moon and, to a lesser extent, the sun. For people living on or near the ocean, all three types of waves are important. As shown in Table 7, the wave period clearly distinguishes these different types of waves. We now will discuss each of these categories of waves in greater detail.

Table 7. Waves classified by their period

Period	Wave Type
Fractions of 1 sec	Ripples
1–4 sec	Wind chop
5–12 sec	Fully developed sea
6–13 sec	Swell
1–3 minutes	Surf beat
10–20 minutes	Tsunamis
12 or 24 hours	Tides

The many different sizes and types of wind-generated waves shown in Table 7 originally were named by sailors. We can best understand this whole scheme of wind-driven waves by following their development during a storm. We start by imagining a day when there is no wind and the sea is perfectly calm. Then, the wind starts blowing. As the air blows across the water's surface, there is a transfer of energy from wind to water. Little *ripples* form on the surface of the water and move in the direction the air is blowing. As the wind continues to blow, more and more energy is transferred, causing the waves to get bigger. During this process, the waves pass through a condition sailors call *wind chop* and finally become a *fully developed sea*. The way energy is transferred from wind to water wave is not fully understood. However, we do know the size of the waves produced depends upon three things. The first two things are the length of time the wind has been blowing and the distance over which the wind has blown. (This distance is called the *fetch*.) The third and most important factor is the wind speed. For every wind speed, there is a maximum size wave that can be produced. The wind may have to blow for several days over a distance of several hundred kilometers before this maximum is reached. But once it is reached, the wave size will not increase, no matter how long the wind blows or how great the fetch is.

In the area where the wind is blowing, the sea is very rough. There are little waves on top of the big waves, and ripples run across the tops of the little waves. White caps appear as the wind seems to blow the tops off the waves. Another way of understanding the occurrence of white caps in this area is to note that as waves grow in height, they become unstable and break. We define the wave *steepness* as the wave height divided by the wavelength.

$$\text{steepness} = \frac{H}{\lambda} \qquad (7.1)$$

A wave in deep water typically becomes unstable and breaks when its steepness becomes greater than $\frac{1}{7}$. Figure 7.1 shows waves with two different steepnesses. In picture A, the waves have a height of 3 meters and a wavelength of 30 meters, so the steepness is $\frac{1}{10}$, and the wave is classified as stable. The wave in picture B, however, is just on the verge of breaking. Although its wavelength is longer (35 meters), its 5-meter height is also greater, so it has a larger steepness of $\frac{1}{7}$. This breaking of steep waves is the primary way in which waves become longer as the wind continues to blow. That is, as the energy is transferred

Figure 7.1. The steepness of deep water waves

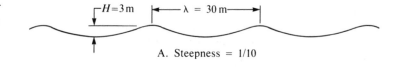

A. Steepness = 1/10

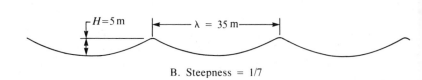

B. Steepness = 1/7

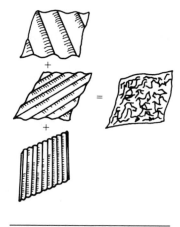

Figure 7.2. Superposition of water waves. Adapted with permission from F.G. Walton Smith, *The Seas in Motion* (New York: Thomas Y. Crowell Co., 1973).

from the wind into the waves, they first become higher and steeper. Then, through this process of breaking, longer wavelength waves are produced.

Once the waves have been generated within the region where the wind is blowing, they move out of that region and can travel for thousands of kilometers across the ocean. Scientists actually have followed waves from storms in the South Pacific near Antarctica, all the way to the coast of Alaska, which is a distance of over 10,000 kilometers. Since these are deep water waves, dispersion is present; as the waves move away from the area of origin, the long wavelength waves travel faster than the short ones. After a trip of several thousand kilometers, the long wavelength waves may arrive a day or two ahead of the short wavelength ones. Once the waves move out of the storm region and dispersion occurs, they take on a smoother pattern. These smoother waves that have a long wavelength are what sailors call *swell*. In addition, the waves lose some energy: Their height becomes smaller. Most of this energy loss occurs near the beginning of the wave's journey, when there is still a great deal of choppiness present. The waves give up the last of their energy when they break against a coast.

When we look at the surface of the ocean, we sometimes can see nice, smooth, regular waves. At other times, such regular behavior is difficult or impossible to perceive. These differences exist because the principle of superposition works here also. In the case of a complicated-looking surface, waves of different wavelengths and different directions of travel all are arriving at the same time. Each of these waves may have originated in a different part of the ocean and is traveling in just the same way as the waves we discussed above. Now, however, the displacement of the surface of the ocean is the sum of the displacements that would be caused by each individual wave. Figure 7.2 is an attempt to illustrate this type of situation. It shows three waves with different wavelengths, wave heights, and directions of travel that are added together (superposed) to give the actual surface. Since this is a still picture, you must imagine how complicated the motion of the ocean's surface will be as the waves pass through this area. Remember that when these are deep water waves, the differing wavelengths will have different wave speeds also.

As mentioned earlier, the ultimate size of wind-generated waves depends upon the wind speed. Typical wave sizes for various wind speeds are shown in Table 8. We must remember, however, that as we move down this table, the wind must blow for a longer period of time in order for the wave to reach the ultimate size listed. It is also important to remember that the wave sizes listed are averages. For each wind speed, waves will be generated that are both larger and smaller than the values listed in the table. This phenomenon is graphically illustrated when we consider the interesting tales that have been told about great storm waves. While many of these stories are undoubtedly exaggerations, there are cases on record that must be taken seriously because of the details involved. One of these cases occurred in 1933 when Lt. (j.g.) F.C. Margraff reported the following fact: While standing watch on the bridge of the USS Ramapo (a Navy tanker 478 feet long), the sea astern appeared above the crow's nest. Using the dimensions of the ship, the calculated minimum wave height was 112 feet. (This is about as high as an 8-story building!)[1]

Waves created by underwater disturbances act very differently from wind-generated waves. Even though they have nothing to do with tides, these waves were called tidal waves for

Table 8. Sizes of waves produced for various wind speeds

Wind Speed (Knots)	Average Wave Height (Feet)	Average of the Highest Ten Percent of Waves (Feet)
10	0.9	1.8
15	2.5	5
20	5	10
25	9	18
30	14	28
40	28	57
50	48	99

Data taken with permission from F.G. Walton Smith, *The Seas in Motion* (New York: Thomas Y. Crowell Co., 1973).

1. From Willard Bascom, *Waves and Beaches* (Garden City, N.Y.: Anchor Press/ Doubleday, 1980, 56–57.

many years. Scientists decided to use the Japanese word for these waves, *tsunami*, in order to minimize any confusion about their origins. For the creation of a typical tsunami, an earthquake causes the sea floor to drop (or rise). The water above the earthquake then vibrates as it tries to return to the equilibrium position. As a result, a wave moves out from the position of the earthquake with crests occurring approximately 15 minutes apart. Thus, the period of the wave produced is about 15 minutes. We find this type of wave travels at a speed of about 200 meters per second, which is about 450 miles per hour. Thus, it has a wavelength of about 180 kilometers (or 110 miles).

The deepest part of any ocean is the Trieste Deep of the Mariana Trench near Guam Island in the western Pacific. The depth there is 35,800 feet, which is less than 7 miles. Thus, even above this oceanic abyss, the wavelength of a tsunami is greater than 14 times the water depth. At all other points in the oceans, the wavelength-to-depth ratio is even larger. Thus, we safely can conclude that a tsunami is a very shallow water wave, and its speed is proportional to the square root of the water depth. A ship in the open ocean hardly would notice as a tsunami passed underneath it because the tsunami's wavelength is so large, and the wave height is less than a meter. However, as the tsunami moves into shallower water near the coast, it starts slowing down. The front end of the wave slows down first, allowing the rest of the wave to catch up and pile on top of it. The result can be

An artist's depiction of a tsunami. This print by the early nineteenth century Japanese artist Katsushika Hokusai is entitled "The Great Wave at Kanagawa." The mountain in the background is Mount Fujiyama. (The Metropolitan Museum of Art, The Howard Mansfield Collection, Rogers Fund, 1936. All rights reserved. The Metropolitan Museum of Art.)

Figure 7.3. Bulges of the ocean producing tides

a large, destructive, breaking wave. This destructive behavior is, of course, what makes the tsunami infamous.

The third category of wave is the *tide*. The tide is created by the gravitational attraction of the moon; to a lesser extent, of the sun; and, to a much lesser extent, the other planets. In order to understand how this gravitational attraction works, let's first consider only the effect of the moon. Figure 7.3 shows the system of the earth and the moon, which rotates around the earth. The elliptical gray section around the earth is an exaggerated view of the ocean. As shown, the ocean bulges on two sides of the earth. On the side nearest the moon, the water bulges because of the moon's gravitational attraction. The bulge on the opposite side of the earth is produced in a different way. The moon actually does not revolve about the fixed center of the earth. Instead, the earth and moon both revolve around a common central point, such as the point marked C in Figure 7.3.* As the earth revolves about this central point, water on the far side of the earth moves away from the center in much the same way that water stays in the bottom of a pail when it is swung in a circle.

The earth and moon complete a revolution about their common center in approximately 27 days. During this time, the bulges of ocean always point toward and away from the moon. The earth turns on its axis once every 24 hours, moving beneath the distorted ocean. Thus, during this 24-hour period, a point on the earth experiences two high tides, one for each bulge in the ocean's surface. From this description, we see that the tide represents a wave whose wavelength is half the circumference of the earth (the distance from the top of one bulge or crest to the other). The period—the time between successive bulges (or crests)—is actually slightly longer than 12 hours because the revolving moon moves the bulges while the earth rotates.

The sun also produces bulges on the ocean's surface. These bulges are much smaller, however, because the sun is much farther away. As a result, we notice the sun's effect on the tides

*The point C is actually inside the earth but has been moved outside in Figure 7.3 for clarity.

mainly as it adds to the effect produced by the moon. The effects add when the sun, earth, and moon all lie along the same line in space. In this case, tides are bigger than normal and are called *spring tides*. Smaller than normal tides, called *neap tides*, occur when a line through the earth and sun is at right angles with a line through the earth and moon. These situations are shown in Figure 7.4. In principle, every other object in the sky produces an effect on the tides. These effects are so small, however, that we conveniently can neglect them.

Surf Beat

As we discussed earlier, the principle of superposition applies to water waves. One effect that the superposing of two sets of water waves can produce is known as *surf beat*. This effect is just another example of beats, which we discussed in Chapter 2. There, we found that the superposition of two vibrations having frequencies f_1 and f_2 gave a vibration having a frequency equal to the average of f_1 and f_2 and an amplitude that varies at a rate equal to the difference of f_1 and f_2.

The same thing happens when we have two sets of water waves with slightly different frequencies arriving at the same location simultaneously. The two original sets of waves might be generated in very different parts of the ocean. On a Carolina beach, for example, one set of waves might have been generated by a storm near Great Britain and the other by a storm off the west coast of Africa. Then, at the beach, when these two sets of waves are in phase with each other, the resulting waves are very large. A short time later, the waves are out of phase, and the water is much calmer.

Figure 7.4. Location of earth, moon, and sun for spring and neap tides (The sizes are not to scale.)

A. Spring tides

B. Neap tides

The old surfer's adage to "wait for the seventh wave" is undoubtedly based on the surf beat effect. However, the seventh wave is not always the largest; the size of the wave depends on the frequencies of the two waves creating the beat. For example, suppose the periods of the two original sets of waves are 10 sec and 11 sec, respectively. Then (by taking the reciprocals), the frequencies of the two waves are $\frac{1}{10}$ Hz and $\frac{1}{11}$ Hz. Subtracting and averaging, we get a beat frequency of $\frac{1}{110}$ Hz and an average frequency of approximately $\frac{1}{10.5}$ Hz. Taking reciprocals again, we find the waves arriving at the beach have a period of about 10.5 sec, and the period of the beats (or the time interval between large waves) is 110 sec. Thus, in this case, every tenth wave is a large one.

The Breaking of Waves

For all types of water waves, there is a transfer of energy to the waves from an outside force. In the cases we have discussed, this energy comes from, respectively, the wind, a movement of the sea bottom, and the gravitational attraction of the moon or sun. Once formed, the water wave carries this energy across the surface of the ocean. Although there is some energy loss during its travel, most of the wave's energy is released when it breaks against the coast.

Two extreme types of breaking waves can occur, depending upon the topography of the ocean floor near the shore. In one case, the floor has a steep slope, so the wave slows rapidly while its crest continues to move at the original speed. As a result, the crest curls over and drops into the preceding trough. We call these *plunging breakers*, and there is a rapid release of the wave's energy so that violent water movements may occur. Swimming or surfing in these types of waves is tiring at best and can be dangerous; you should exercise caution before you rush into this type of surf.

Waves breaking where the depth of the water changes gradually are in stark contrast to these plunging breakers. Here, the waves slowly build up in height, tip gently forward, and roll onto their front faces. For obvious reasons, we call these *spilling breakers*. To see why these waves behave differently from plunging breakers, let's look at how and why spilling breakers are formed. When the depth of the water decreases to one-half the wavelength, the wave starts to feel the bottom, which means the wave is changing its character from a deep water wave to a shallow water wave. As mentioned earlier, the speed of shallow water waves depends on the depth of the water. As the depth of

Plunging breakers, which are common at a beach when there are large waves (Photograph courtesy of Steve Lissau.)

the water continues to gradually decrease, the waves slow down. (An easy way of remembering this is to think of the waves as dragging on the bottom.) Something else happens when a wave starts to slow down: Its height increases. This behavior is reasonable because the total amount of water in the wave tries to stay the same; when the wave gets shorter, it also gets taller. As this process continues, the height-to-length ratio grows (the wave becomes steeper), so the wave becomes unstable and breaks.

Scientists have found that these gradually changing waves break when their height is about three-fourths the depth of the water. An example of this behavior is shown in Figure 7.5. In this figure, the deep water waves far from shore have a wave-

Spilling breakers, which are more gentle than plunging breakers (Photo courtesy of Steve Lissau.)

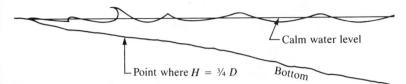

Figure 7.5. Waves breaking on a gently sloping beach

length equal to 20 feet and a height equal to 2 feet. Thus, when the water depth becomes 10 feet, the waves start slowing down, thereby, decreasing their wavelength and increasing their height. Farther toward shore, the depth has decreased to 4 feet, and the wave height has increased to 3 feet, so the wave there is starting to break.

Mass Transport

We mentioned earlier that particles of water move on circular or elliptical paths as waves pass. It is implicit in this description that the water moves back to its original location after each wave cycle. That is, the waves cause no migration of the water across the surface of the ocean. This description is approximately true for non-breaking waves, but it is not true for waves that break. As can be seen easily on a beach, breaking waves move water up on the sand, and the water then runs back into the ocean. Any material that is suspended or floating in the water— such as sticks or small particles of sand—is carried along with the water as the wave breaks. This process, whereby water and suspended material is moved from one place to another, is termed *mass transport*. The movement can occur either in a direction parallel to the shore (which we call *longshore transport*) or perpendicular to the shore.

We observe longshore transport when the waves arrive at an angle to the shore. As the waves break, of course, they carry water and suspended material with them. But since they break at an angle, the water and material is moved at an angle up onto the beach. Then, the water runs almost straight back down, so it enters a new breaking wave at a point farther along the beach. We can think of this behavior, which is shown in Figure 7.6, as being similar to a stream or river transporting water and material slowly along the shore. This behavior happens primarily within the region where the waves are breaking, so there are boundaries on each side of this imaginary river. Although the river is imaginary, there is definitely migration of water and material; this migration sometimes is called the *longshore current*. In 1950, F.P. Shepard observed a dramatic example of this longshore transport. During a single day, waves moved sand

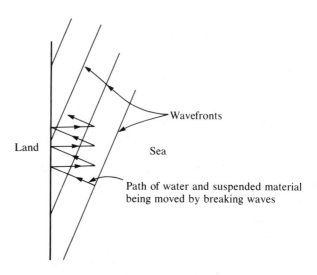

Figure 7.6. An overhead view of breaking waves showing longshore transport

from one end of Boomer beach (which lies in a restricted bay on the California coast) and deposited the sand at the other end. The level of the beach receiving the sand was raised by 2.44 meters (about 8 feet), and this happened in less than 24 hours![2]

When waves strike the shore head-on, mass transport also takes place. In this case, however, sand and rock, which are carried along with the water, either build the beach up or tear it down. This process is very complicated. The amount of material moved depends on many factors. Some examples of these factors are

1. the size of the rocks or sand particles

2. the bottom's slope and character (that is, whether the bottom is rippled or smooth)

3. the length and height of the waves

4. the wave steepness

5. the size of the tides

6. the water depth

One of the complications in this process is that there is some mass transport beyond the breaking waves. This additional mass transport occurs because water waves are asymmetrical in

2. *F.P. Shepard, Beach Cycles in Southern California.* B. & B. Technical Memo 20 (1950). Taken from Cuchlaine A.M. King, *Beaches and Coasts*, 2nd ed. (New York: St. Martin's Press, 1972) 420.

shape, so the speed of the water is faster as it moves forward with the wave than when it moves back after a crest passes. This condition is important, of course, only for shallow water waves where the water at the bottom moves horizontally. The faster moving water stirs up the bottom material and carries it forward. The returning, slower water creates a much smaller disturbance, so it moves less material back. The net result is transport of bottom material toward the shore in the region beyond where the waves are breaking.

Within the region of breaking waves, material may be moved either toward or away from the shore. In the former case, we have *accretion* of new material and a building-up of the beach. The opposite effect of *erosion* occurs when beach material is moved out to sea. The most important factor in determining whether breaking waves will produce accretion or erosion is the wave steepness. Small, long wavelength waves move material toward the shore. As the steepness increases, a break-even point is reached where material is not moved in either direction. For even steeper waves, erosion occurs. The critical value of wave steepness at the break-even point seems to change with the size of the rocks or sand being moved. Some scientists have found that the critical steepness increases as the size of the material increases.

On the west coast of the United States, we find an interesting example of mass transport. Since this coast is exposed to waves that are generated in almost any part of the Pacific Ocean, it experiences a wide variety of wave sizes and steepnesses. In particular, storm waves arriving during winter are steeper than summer waves. Summer waves are smaller for two reasons. First, summer storms are less severe than winter ones. Therefore, in the North Pacific, the summer storms produce smaller waves than do the storms in winter. Of course, during California's summer, there are winter storms in the South Pacific. But we have already seen that waves moving from the South Pacific to the California and Oregon coasts lose some energy and height. Since summer waves are less steep than winter storm waves, we find a seasonal variation of the wave steepness. This seasonal variation produces an annual cycle of sand movement. During the summer, the small waves carry sand inward and rebuild the beach that was eroded the previous winter. Then, the following winter, the large waves pull the sand off the shore again and move the sand out under the water. Thus, by springtime, the actual beach is much smaller than it was in the autumn. But after a full year, the beach is back to almost where it started.

Funneling Effects

We have seen several examples of how the ocean bottom can affect the water movement and wave behavior. Two other examples are rip currents and large waves (created by tides or tsunamis). They are related because in each, the water is forced into a narrow channel by the geography. In another sense, however, these examples contrast because the first involves water returning to the ocean following the breaking of waves while the second is concerned with incoming waves.

When you stand in surf, you can feel the water running back to the ocean after it was carried up the shore by breaking waves. As the water pulls on your feet, it often feels like it will suck you out to sea. Indeed, there is a commonly held belief that this *undertow* will pull people far out to sea. The true situation is much less dangerous, however, because the returning water producing the undertow goes no farther out to sea than where the waves are breaking. There, the water becomes part of the next wave.

In contrast, a *rip current* is extremely dangerous; it can carry an unsuspecting swimmer far out to sea almost before he realizes what is happening. Fortunately, rip currents are far less common than the ordinary undertow, which occurs whenever there are breaking waves. In order to have a rip current, there must be a sandbar or other geographical obstruction that can trap water near the shore. The trapped water then runs back to the sea through openings in the obstruction that act as channels. Such a situation is pictured in the map of Figure 7.7, where the dashed lines show submerged sandbars. The waves breaking in the shallow water over the sandbars will carry a large amount of water into the region between the bars and the shoreline. As this water tries to return to the sea, some of it will pass directly over the bars, but much of it will run through the opening in the bar, creating a current. With the proper conditions, this rip current can flow much faster than a person can swim. If you ever are caught in such a rip current, you should swim across the current (that is, parallel to the shore) rather than trying to swim directly toward shore. Then, when you are out of the current, you can turn toward shore.

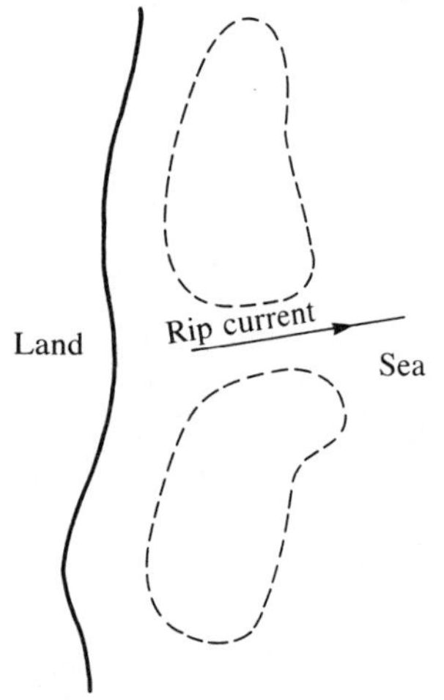

These ideas about ocean waves are difficult to demonstrate in a laboratory. However, they are interesting effects which you can look for on your next journey to a large body of water. Surf beat is quite common, and knowing why the wave height changes in a periodic way can be very satisfying. Watching the waves change direction as they approach the shore is an excel-

Figure 7.7. A map showing submerged sandbars and a channel for a rip current

lent demonstration of refraction. For many people, there is an almost hypnotic fascination with watching waves roll into the beach. This fascination is probably due to the endless variation for which the sea is famous. The waves have regularities, however, and you now can appreciate them. For our purposes, though, waves moving on water can be seen, and the same behavior we see there will show up in both sound and light where we cannot see the waves. Before we look at how such concepts as refraction show up with light and sound, we first will look at some additional ways that waves behave. In the next chapter, we discuss standing waves, which will help to explain how various normal mode vibrations occur in strings and other complicated vibrating objects.

Summary

The majority of ocean waves are generated by wind blowing across the surface of the water. The size of the waves produced depends upon three factors: the speed, duration, and fetch of the wind. Waves created in one part of the ocean can travel thousands of kilometers across the sea. They lose a small part of their energy during the trip; most of it is released when they crash against the shore. Other types of waves are tsunamis and tides, which are generated by underwater disturbances and the gravitational pull of the moon and sun, respectively. Non-breaking water waves satisfy the principle of superposition, which helps us understand why the water's surface is sometimes very irregular and how surf beats are produced. There are a number of ways that waves break, depending on the shape of the ocean bottom. Two noteworthy examples are plunging breakers and spilling breakers. The commonly held belief about undertow being dangerous is not true, but rip currents can be treacherous.

Suggested Readings

The book *Waves and Beaches,* by Willard Bascom (Garden City, N.Y.: Anchor Press/Doubleday, 1980) has many fascinating stories connected with the sea. Bascom describes his own work in the study of beach formation and erosion. He has many interesting stories of great waves, and he describes in some detail how scientists measure wave properties and behavior. The level of the book is about the same as this book's level.

Ocean waves and their effects on coastal areas usually are discussed in oceanography books. One that is quite complete is *Elements of Oceanography,* by J. Michael McCormick, and John V. Thiruvathukal (Philadelphia: W. B. Saunders Company, 1976). The book is written at the same level as this book, although *Elements of Oceanography* contains more details.

A number of different tsunamis that have occurred throughout history are described in the article, "Tsunamis." (*Scientific American,* August 1954.) In addition, methods of detection and warning are discussed.

The January-February 1979 issue of *Oceans* has some beautiful pictures in a photography essay entitled "The Breaking Wave," by Warren Bolster. The issue also has a nice article about predicting wave heights entitled "High Surf in Hawaii," by Steve Lissau.

Review Questions

1. Standing on the deck of a ship, you watch waves slapping against its side. You note that in 20 sec, seven waves strike the ship. Therefore, a sailor would probably classify the waves as

   a. ripples.

   b. wind chop.

   c. a fully developed sea.

   d. swell.

   e. a surf beat.

2. In the open ocean, waves 3 meters high are just beginning to break. What is the wavelength of these waves?

   a. 3 meters

   b. 4 meters

   c. 12 meters

   d. 21 meters

   e. 35 meters

3. Waves in the open ocean have a period of 8 sec and a speed of 12.5 m/s. How high will these waves be when they start to break?

   a. 1.8 meters

   b. 4.5 meters

   c. 10.9 meters

   d. 14.3 meters

   e. 56.0 meters

4. In 1946, an underwater earthquake occurred near the coast of Alaska. About how long did it take for the resulting tsunami to reach Hilo, Hawaii, which is about 2000 miles to the south?

   a. 5 hours

   b. 20 hours

   c. 2 days

   d. 5 days

   e. 10 days

5. A group of scientists are following some waves generated in the South Pacific all the way to the Alaskan coast. The longest waves they watch have a wavelength of 225 meters and a speed of 18.7 meters per second. The shortest waves have $\lambda = 60$ meters and $v = 9.7$ m/s. The total distance is 10,000 kilometers (10^7 meters).

   A. About how long does it take for the first wave to make the trip?
      a. 8.6×10^4 sec (one day)
      b. 18.7×10^4 sec (2.2 days)
      c. 5.34×10^5 sec (6.2 days)
      d. 9.61×10^5 sec (11.1 days)
      e. 17.6×10^5 sec (20.4 days)

   B. When the first wave hits the coast, about how far behind is the last one?
      a. Several meters
      b. Several hundred meters
      c. Several thousand meters
      d. Almost halfway across the ocean

6. Waves approaching the shore over a gently sloping bottom have a wavelength of 30 meters and a height of 2 meters. If the wave height did not change, at what water depth would the waves break?

   a. $2/7$ meters

   b. $1\frac{1}{2}$ meters

   c. $2\frac{2}{3}$ meters

     d. $4\frac{2}{7}$ meters

     e. 8 meters

7. You arrive at an unfamiliar beach and observe that 4-meter-high waves arriving every 15 sec curl over and pound against the water in front. What can you conclude about the bottom?

     a. The bottom is gently sloping and is 3 meters deep where the waves are breaking.

     b. The bottom is gently sloping and is $5\frac{1}{3}$ meters deep where the waves are breaking.

     c. The bottom is steeply sloping and is 3 meters deep where the waves are breaking.

     d. The bottom is steeply sloping and is $5\frac{1}{3}$ meters deep where the waves are breaking.

8. Two waves created in different parts of the Pacific arrive at San Diego, California. One has a period of 15 sec, and the other has a period of 18 sec. How many seconds will occur between times of big waves?

     a. 3

     b. 30

     c. 33

     d. 60

     e. 90

9. The moon circles the earth once every 28 days (approximately). A spring tide occurs on July 2. When will the next neap tide occur?

     a. July 6

     b. July 9

     c. July 16

     d. July 23

     e. July 30

10. You arrive at an unfamiliar beach; a posted sign reads "Heavy Surf Today—Beware of the Dangerous Undertow." How should you react to this warning?

a. Ignore it because undertows are not dangerous.

b. Be careful because undertows are dangerous.

c. Ignore it because undertows do not occur when there is heavy surf.

d. Be careful because even though undertows are not dangerous, there may be rip currents, which can be dangerous.

8

Standing Waves

Important Concepts

- A wave on a string is reflected with inversion (or has a 180° phase change) at a fixed end. No inversion occurs for reflection from a free end. In each case, the amplitude of the reflected wave is the same as that of the incident wave.

- A wave on a string reflected from a dashpot or the junction with another string may have a smaller amplitude and be either upright or inverted.

- In general, when a wave on a string encounters the junction with another string where wave speed is different, there will be both a reflected and a transmitted wave.

- Whenever two waves having the same frequency and amplitude travel in opposite directions through the same region, they will superpose to produce a standing wave.

- Standing waves in a finite-length system such as a string or a pipe are the system's normal modes of vibration. The frequencies and wavelengths of the waves depend on the types of ends, the wave speed, and the length of the string.

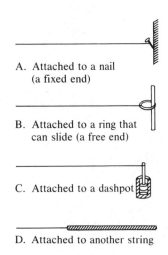

A. Attached to a nail
   (a fixed end)

B. Attached to a ring that
   can slide (a free end)

C. Attached to a dashpot

D. Attached to another string

Figure 8.1. Four ways to end a string

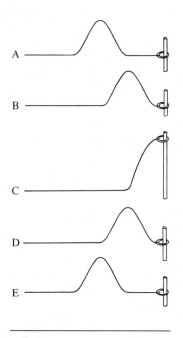

A

B

C

D

E

Figure 8.2. Reflection of a crest from a free end

- The frequencies of the fundamental modes and the possible harmonics for the two possible sets of end conditions are

| Ends alike | $f_1 = v/2L$ | All harmonics |
| Ends different | $f_1 = v/4L$ | Only odd harmonics |

- For a string, there is always a node at a fixed end and an antinode at a free end.

- For a pipe, there is always a displacement node and pressure antinode at a closed end and a displacement antinode and pressure node at an open end.

- The wavelength of the traveling wave that produces the standing wave is twice the length of a loop of the standing wave, that is, twice the distance between adjacent nodes.

In Chapter 6, we considered the movements of a transverse wave on a string and a longitudinal wave on a slinky. The medium through which each wave traveled was, respectively, the string and the slinky. In our discussion, we considered only the situation where the medium did not change—or end. In this chapter, we will look at cases in which the medium changes; we will find that reflection of waves can occur. We will consider waves on the string and slinky first, and we will show how the superposition of the reflected wave and the original or *incident* wave can produce standing waves. Then, we will look at other examples of standing waves such as sound waves in pipes and water waves in enclosed areas. In Chapter 9, we will use the knowledge gained in this chapter to help us understand how musical instruments work.

Wave Reflection at the End of a String

First, let's look at what happens when a transverse wave traveling along a string reaches the end of the string. Through experimental observation, we find that the way the end of the string is attached produces different effects. What possibilities exist? First, the end of the string could be tied to something rigid such as a nail driven in a wall as shown in Figure 8.1A. In this case, the end of the string cannot move; we term it a *fixed end* of a string. Figure 8.1B shows a second way of attaching the end of a string. There, we see the end is tied to a ring that can move (without friction) along a rod perpendicular to the string. This method of connection allows the string tension to be maintained while the end of the string is free to move in a direction transverse to the string. We term this a *free end* of a string. A third way of terminating a string is to attach it to a device we call a *dashpot*. This mechanism allows the end of the string to move in a transverse fashion, but the end moves with friction. This method of attaching a string end is shown in Figure 8.1C. We also could end a string by attaching it to another string as shown in Figure 8.1D. Each of these ways of ending a string will produce a different result when a transverse wave arrives at that end. Thus, we will investigate each one of them in turn.

In order to see what happens to a wave, let's consider individual pulses arriving at the end of the string. First, we will look at a free end because it is the easiest to visualize. The sequence of pictures in Figure 8.2 shows a single crest arriving at the end of the string and a new (reflected) pulse moving back along the string in the opposite direction. As shown in picture C, when the crest arrives at the end of the string, the ring to which the string is attached moves vertically upward. As shown, when the crest reaches the ring, it moves higher than elsewhere on the string. This extra movement occurs because of the forces involved. Away from the end, the string on both sides of the crest is pulling it down, but at the end, there is no string on the far side of the ring. This extra displacement of the ring acts on the string in the same way that the original end did when it was pulled upward. (See Figure 6.1 in Chapter 6.) That is, it causes a new crest to move backward along the string in the opposite direction. Thus, a crest arriving at the end of the string produces a crest moving in the opposite direction.

Another way of describing this effect is to say the crest is reflected at the end of a string. There is also a reflection when a crest arrives at a fixed end of a string. In this case, however, the crest is reflected as a trough. The reason a trough returns along the string when a crest moves into the string's end is connected with the forces involved also. As shown in Figure 8.3, the crest arriving at the string's end cannot displace the end. (That's what we mean by a fixed end!) Even though the crest does not move the string's end up, it still pulls upward on the nail (or whatever the string is fastened to). But as Sir Isaac Newton first pointed out three centuries ago, when the string pulls up on the nail, the nail pulls down on the string. (This phenomenon is an example of Newton's famous third law of motion, which is often stated: "For every action, there is an equal and opposite reaction.") Since the nail pulls down on the string, a trough is produced. This trough has only one direction to travel—back along the string. As before, we can describe this effect in terms of reflec-

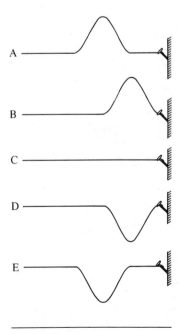

Figure 8.3. Reflection of a crest from a fixed end

tion: A crest arrives at the fixed end and is reflected as a trough. We say there is *inversion* when the pulse (or a wave) is reflected from a fixed end of a string. In both cases of the free end and the fixed end, the reflected pulse has the same amplitude as the incoming pulse.

A wave, of course, is just a series of crests and troughs following one another along the string. Thus, it is easy to see what happens to a wave reflected from a fixed or a free end of a string. First, we consider only one cycle of a wave as consisting of one crest followed by one trough. When this wave reaches a free end of the string, the crest arrives first and is reflected as a crest. The trailing trough is reflected as a trough and still trails the crest in the reflected wave. Thus, for a free end, the reflected wave is the same as the incident wave (crest followed by trough). In contrast, a fixed end on the string will invert both the crest and the trough. Consequently, in this case, an incident wave consisting of a crest followed by a trough will produce a reflected wave that is a trough followed by a crest. We also describe these reflections in terms of the phase of the reflected wave compared to that of the incident wave. For reflection from a fixed end of a string, the wave changes its phase by 180°. There is no phase change, however, when reflection occurs at a free end of a string. Just as in the case of pulses being reflected from fixed or free ends, the amplitude of the reflected wave is the same as the incident wave's amplitude.

Before describing what happens to a wave arriving at the end of a string fixed to a dashpot, let's first discuss the dashpot, itself, in more detail. Basically, a dashpot is any device that will absorb energy from the wave. An easy way to construct a dashpot is to have a piston moving through a fluid-filled chamber. This is the type of dashpot illustrated in Figure 8.1C. When a wave arrives at the end of the string attached to such a device, the wave causes the string to move up and down. The string pulls the piston up and down through the fluid, and the viscous drag of the fluid on the piston causes some of the energy of the wave to be transferred through the piston to the fluid. By changing the size of the piston or by changing the fluid, the viscous drag can be adjusted. If the piston moves through a very viscous fluid (such as molasses, for example), then the end of the string will move very little; it is almost a fixed end. If, on the other hand, the piston moves through air (a fluid having a very low viscous drag), then the end of the string will act almost as a free end. Thus, by using a dashpot with various fluids, it is possible to change the way the end of the string affects a pulse or wave. It is important to recognize that we can make small changes in a dashpot so that we can see how reflection varies when we change from a fixed end to a free end.

Now, we will consider what happens to waves reaching the end of a string that is attached to a dashpot. If the dashpot is very stiff (so that the end of the string is similar to a fixed end), there will be a reflected wave that is inverted. As we reduce the stiffness of the dashpot, we will continue to get an inverted, reflected wave, but, as shown in Figure 8.4 (for a single pulse), the inverted, reflected wave's amplitude will be smaller than the amplitude of the incident wave. (Some of the incident wave's energy is absorbed by the dashpot; the reflected wave has less energy and, therefore, a smaller amplitude.) As the stiffness of the dashpot is reduced even further, we eventually will obtain just the correct viscous drag for which no reflected wave is generated. In this case, all of the energy being carried by the wave to the end of the string is absorbed by the dashpot. A continued reduction in the stiffness of the dashpot will produce a reflected wave that is not inverted. Unlike the case of the free end of a string, however, this reflected wave will have a smaller amplitude than the incoming wave. The dashpot is absorbing some of the incident wave's energy. Finally, when the drag of the dashpot has been reduced to zero (so the string has a free end), the reflected wave will have the same amplitude as the incoming wave. Thus, by terminating a string with an adjustable dashpot, we can produce a wide variety of reflected waves.

The last way of ending a string—that is, by attaching it to another string—is very similar to the case of the dashpot. Attaching a string to another string is very similar to the case of the dashpot because the amplitude and phase of the reflected wave (compared to the incident wave) depend on the relative sizes of the two strings. If the second string is larger than the first, the reflected wave will behave as it would when bouncing off a stiff dashpot: The reflected wave will be inverted. On the other hand, if the second string is smaller than the first, the situation is similar to a loose dashpot, and the reflected wave is not inverted. In either case, the reflected wave will have a smaller amplitude than the incoming wave. As in the dashpot case, the incident wave carries more energy than the reflected wave. In contrast to the case of using a dashpot, however, the rest of the energy is carried away by another wave that continues on down the second string. The speed of this *transmitted* wave on the second string is different than the speed of the incident and reflected waves on the first string. This difference in

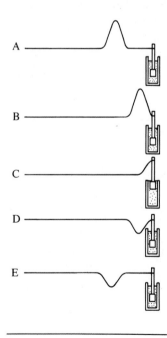

Figure 8.4. Reflection of a crest from a dashpot

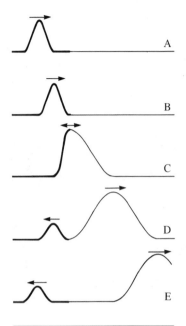

Figure 8.5. Reflection and transmission of a crest from a junction of two strings

speed exists because the medium has changed, and the medium determines the speed of the wave. However, all three waves have the same frequency. Both the reflected and transmitted waves are produced by the incident wave arriving at the junction of the two strings. Thus, the vibration rate for each of the new waves is determined by the frequency of the incident wave. Since the wave transmitted along the second string has a different speed and the same frequency as the incident wave, the wave on the second string will have a different wavelength. The transmitted wave moves faster and, therefore, has a longer wavelength when the second string is lighter than the first.

The arrival of a crest that is incident on the junction of two strings and the generation of the reflected and transmitted pulses is shown in Figure 8.5. Since the second string is lighter than the first, the reflected and transmitted pulses will be crests. Also, the wave speed in the second string is higher, so the transmitted crest is longer than the incident or reflected crests. Finally, the transmitted pulse has a larger height than the incident pulse. This difference in height is also due to the fact that the second string is lighter. The amount of energy carried by a pulse depends on both the density of the string and the height of the pulse. Thus, the smaller incident pulse has more energy than the larger transmitted pulse because the first string has a larger density.

The situation just described is very typical for waves. That is, when the speed of a wave changes abruptly, a reflected wave is usually produced. Also, whenever a reflected or transmitted wave is produced, it will have the same frequency as the incident wave. Since the reflected wave will be in the same medium as the incident wave, the reflected wave will have the same speed and wavelength as the incident wave. The different speed of the transmitted wave will produce a different wavelength also. In addition to waves on strings, we will find these conditions happen for waves on slinkies, sound waves, water waves, and light waves.

The Production of Standing Waves

Now that we have seen how waves are reflected when there is a change in the medium, let's look at what happens in that part of the medium where both the incoming wave and the reflected wave are located. (In the case of a string, this location is near its end.) In that part of the string where both waves coexist, we find the displacement of the string is the sum of the displacements caused by each individual wave. In other words, waves on a string satisfy the principle of superposition.

First, we will consider a string with a fixed end. Remember that for this case, the reflected wave is inverted from the incoming wave. The sequence of drawings in Figure 8.6 shows the position of each wave (that is, the incident and the reflected waves) and the resulting displacement of the string at intervals of time equal to one-fourth the period. We start by looking at point P on the string. In picture A, the incident wave is nearing the end of the string; the wave has just reached point P. Then, one-fourth of a cycle later, the incident wave reaches the end of the string, and point P is raised to the top of the crest, which is now at that location. As soon as the incident wave reaches the string's end, the reflected wave starts back along the string.

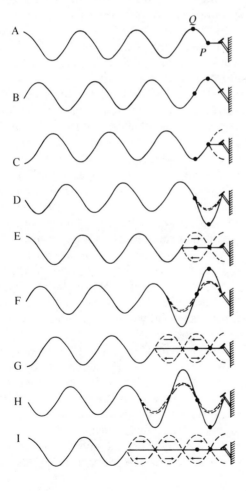

Figure 8.6. The production of a standing wave by superposing the incident and reflected waves near a fixed end of a string

Both the incident and reflected waves are shown as dashed lines where they overlap in pictures C–H. The actual shape of the string is shown as a solid line in all the pictures. Note that in C, the incident and reflected waves only superpose on that part of the string between point P and the fixed end. In each succeeding picture, the length of the string over which superposition occurs increases by one-fourth the wavelength because that is the distance the reflected wave travels in one-fourth the period.

As the incident wave continues to move to the right and as the reflected wave moves to the left, point P executes simple harmonic motion with an amplitude that is equal to twice the amplitude of the incident wave. In contrast, point Q remains stationary. What is the reason for this difference? By following the movement of the incident and reflected waves in pictures D–H, we can understand why P oscillates with a large amplitude while Q remains stationary. Picture D shows a deep trough located at P, and this deep trough occurs because both the incident and the reflected waves have troughs there at that time. In the same picture, Q is not displaced because neither of the waves shows a displacement at that point. One-fourth of a cycle later (picture E), Q still is not displaced from the string's equilibrium position, but now, this lack of displacement occurs for a different reason: The trough of the reflected wave exactly cancels the crest of the incident wave. One of these two situations always happens at point Q. That is, displacement in one direction due to one of the waves always will be canceled by an equal but opposite displacement due to the other wave, or neither wave will produce any displacement at all.

With the superposition of the two traveling waves, the string executes a motion called a *standing wave*. We give the motion this name because there is no apparent motion of the wave along the string. We call the points on the string that do not vibrate during standing wave motion (such as point Q) *nodes* and those points that vibrate a maximum amount (such as point P) *antinodes,* just as we did when describing strings vibrating in various normal modes.

The result we have just found—that is, that standing waves are produced when two waves are traveling in opposite directions through the same medium—is very general. This result occurs for all types of waves. In the case we just discussed, the amplitude of the reflected wave was the same as the amplitude of the original wave. Therefore, the sum of an incident crest and a reflected trough exactly canceled, and the string had no motion at the nodal positions. It is also important to note that because we had a fixed end on the string, there was a node at the string's end. After all, fixing the end of the string prevented it from moving. This node only can occur, of course, if the two traveling waves are 180° out-of-phase at the end of the string. And we saw earlier that a wave reflected from a fixed end has its phase changed by 180°.

For the other ways of terminating the string that were just discussed, we get different types of standing wave patterns. When we have a free end of the string, the reflected wave has the same amplitude as the incident wave and is not inverted. Thus, superposing the incident and reflected waves creates an anti-node at the end of the string. This case is shown in Figure 8.7. I must emphasize two differences between Figures 8.6 and 8.7. First, there is no change of phase for the reflected wave at a free

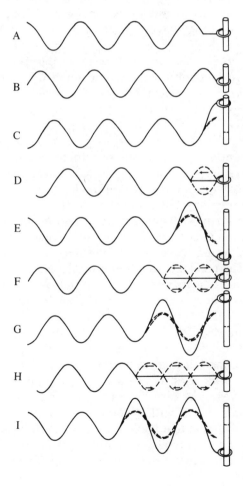

Figure 8.7. The production of a standing wave by superposing the incident and reflected waves near a free end of a string

end (Figure 8.7), while a 180° change occurs at a fixed end (Figure 8.6). These conditions produce the second difference: An antinode occurs at the free end shown in Figure 8.7, compared to the node at the fixed end in Figure 8.6.

Since it will be important later, let me point out a similarity between the standing waves that occur in Figures 8.6 and 8.7. In each, the distance between adjacent antinodes (or between adjacent nodes) is one-half the wavelength of the traveling waves that superpose to form the standing wave. This situation is easily seen in picture H of Figure 8.6 or picture I of Figure 8.7. Both of these pictures show the maximum string displacement occurring where the traveling waves have either a crest or a trough. But since an antinode is located at both the positions of a high crest displacement and a low trough displacement, the distance between antinodes is only one-half the wavelength of the traveling waves. We will want to use this comparison later when we discuss standing waves on strings where both ends are considered.

In the other two ways of terminating the string, the reflected wave does not have the same amplitude as the incident wave. In these cases, superposition of the incident and reflected waves produces only a partial standing wave. In this partial standing wave, the nodal points on the string are not at rest (in the superposition of unequal crests and troughs, they cannot completely cancel). It is sometimes useful to describe the string's movement in such a case as being the superposition of a standing wave and a traveling wave, where the latter moves in the incident wave direction. The amplitude of this traveling wave is equal to the string's displacement at the nodal points of the standing wave. This type of situation is shown in Figure 8.8, where the reflected wave is not inverted, but its amplitude is only half of the incident wave's amplitude. Note that point Q is an antinode because its vibration amplitude is larger than other points. Point P, however, is not a true node because it has some movement.

To summarize: We have found that two traveling waves having the same frequencies and wavelengths and moving in opposite directions through the same medium produce a standing wave. In order to produce a pure standing wave, the two traveling waves must also have the same amplitude. Further, these standing waves will be produced for any frequency or wavelength of an incident traveling wave that is reflected from a fixed or free end of a string. (This finding is in direct contrast to what we will find when we consider a string of a definite length

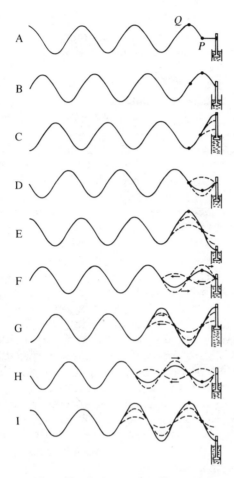

Figure 8.8. The production of a partial standing wave by superposing incident and reflected waves with unequal amplitudes

where we worry about both its ends.) For waves on a stretched string, there will be a nodal point of the standing wave at a fixed end of the string and an antinodal point at a free end.

Normal Modes as Standing Waves

In the previous discussion, we were concerned with only one end of a string, and therefore, the length of the string was immaterial. Now, we wish to consider both ends of the string and to look at how they are attached. In this case, the length of the string, or *L,* will be important. We will find that standing waves occur for only certain values of the wavelength of the traveling waves that produce standing waves. Since the speed of the traveling wave is determined by the medium (in this case, the string), only certain frequencies of vibration will be allowed. There are two ways to see why this effect occurs, and we will look at both of them. The two different explanations complement each other, so that looking at both should help us gain a fuller understanding of this important idea.

For the first way of looking at this situation, let's consider a string that is fixed at each end, but that we can jiggle a little bit at its left end. We jiggle the end and start a crest moving along the string, as shown in picture A of Figure 8.9. When the crest reaches the right end (picture C), it is reflected (with inversion) and moves back toward the left end as a trough (picture D). When it reaches the left end in E, it is inverted again upon reflection and starts back toward the right as a crest. Note that there are two pulses shown as dashed lines in each of the pictures C, E, G, and I. In each case, the two pulses are occurring at the same time, and they superpose to produce the straight string that is shown as a solid line. The single pulse in this figure will move back and forth along the string, as a crest when moving to the right and as a trough when moving to the left. For a real string, the size (amplitude) of the pulse will die out. In order to maintain a standing wave on the string, we must continue to jiggle its end, creating new pulses. In this way of looking at the production of normal modes, we want to recognize that to produce a standing wave, we must put the new pulses on the string in phase with the pulse that exists there already. That restriction means we must put a new crest on the string just when the first crest is being reflected from the left end. In order to put a new pulse on the string every time the first one makes one round trip, we must jiggle the end of the string at a certain frequency or with a certain period. The period is just equal to the time required for the round trip of the original pulse. This time depends upon two things: the length of the string, or L, and the speed of the wave on the string, or v. Solving the simple distance-rate-time relation for the time tells us the period is equal to the distance divided by the speed, or

$$T = \frac{D}{v} = \frac{2L}{v} \qquad (8.1)$$

Thus, the frequency is equal to the speed of the wave divided by twice the length of the string.

$$f = \frac{1}{T} = \frac{v}{2L} \qquad (8.2)$$

We must jiggle the end of the string at just this frequency to produce a standing wave. If we move the string a little faster or a little slower, the jiggles will cancel one another out rather than

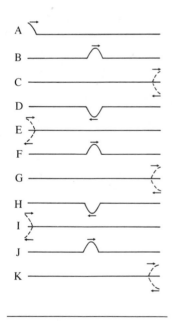

Figure 8.9. The movement of a single pulse on a string with two fixed ends

reinforcing each other. There are, however, some other frequencies for which a standing wave will occur. Let's see why this statement is true. Instead of waiting for the original pulse to make a complete round trip before putting a second pulse on the string, we could put the second pulse on the string when the first one reaches the right end. Then, the third pulse would match up with the original one, the fourth pulse would match up with the second, and so forth. This would just double the frequency at which we put pulses on the string. We also would get a standing wave on the string. However, the frequency would be twice as large, and there would be a node at the center of the string. This situation is shown in Figure 8.10, and we can see the development of the node at the center because the crest moving to the right and the trough moving to the left cancel in pictures G and K. There are two things we should notice about this figure. First, during each round trip of a pulse on the string, there are more pictures shown in Figure 8.10 than in Figure 8.9. Thus, the time interval between adjacent pictures is shorter in Figure 8.10 than in Figure 8.9. The second thing to note is that the nodes at each end of the string are shown in the same way they were shown in Figure 8.9. That is, in picture E, the two pulses (shown with dashed lines) occur at the same time, and the superposition of the incident crest and reflected trough at the right end cancel each other out.

As another possible way of putting pulses on the string, we could have the fourth pulse match up with the first one so there are a second and third in between the two. In this case, the frequency would be three times the frequency used when the second pulse matched up with the first one. This option of putting any number of pulses in between the first one and the one that matches up with it means there are many different frequencies that will produce standing waves on this string. As we can see, these frequencies are all integer multiples of the original frequency. (This statement should sound familiar!)

The pulses that are shown in Figures 8.9 and 8.10 have been drawn in a distorted way in order to show their movement clearly. When such pulses are used to produce standing waves, each pulse is much wider. The pulses in Figures 8.9 and 8.10 should have widths equal to the length of the string and half the length of the string, respectively. This correction is made in Figures 8.11 and 8.12.

Here, we see a great similarity to the normal modes of vibration for an ideal string that we discussed in Chapter 3. This similarity is no accident. The concept of standing waves is just another way of describing normal modes of vibration, but it

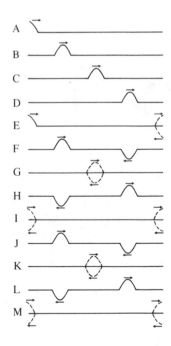

Figure 8.10. The movement of two pulses on a string with two fixed ends

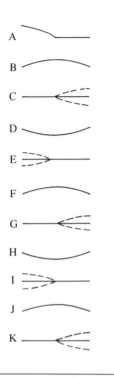

Figure 8.11. The pictures in Figure 8.9, but with pulses that have the correct width

gives us a greater understanding of what is happening. Now, we can see that when a string is made to vibrate—either by hitting, bowing, or plucking—pulses are sent along the string and are reflected from its ends. Those vibrations that are at the correct frequency to cause the original pulses to be reinforced are those that build up. Such frequencies are just the frequencies of the normal modes of vibration.

This discussion allows us to understand the occurrence of harmonics on ideal strings. In order to have harmonics, the frequency of the standing wave produced in Figure 8.10 (or Figure 8.12) must be exactly twice that of the wave of Figure 8.9 (or Figure 8.11). This requirement, in turn, means these waves (of different wavelengths) must travel at the same speed. The reason the normal modes of a stiff (piano) string are not harmonics is because the short wavelength waves travel faster than the long wavelength ones. This higher speed for the shorter wavelengths means the travel time for a round trip is less than for the longer wavelengths. Consequently, the period for the second normal mode, for example, is slightly less than twice the period for the first normal mode. Since this situation occurs for all the higher normal modes as well, they all have frequencies slightly higher than the harmonic values. As pointed out in the last chapter, when different wavelength waves travel at different speeds, we call it dispersion, and we will discuss this concept more fully in Chapter 11.

We also see from this discussion that for an ideal string with two fixed ends, all the harmonics can occur. In contrast, only the odd harmonics can occur on a string with one fixed and one free end. We now will see why this condition is true using the same type of development. We also will find that the fundamental frequency is only one-half that for the above case, even though the string lengths and wave speeds are the same for both cases.

For a string with one end fixed and one end free, we put its fixed end at the left and its free end at the right. As before, we create a crest by jiggling the left end. In Figure 8.13, we show how the (narrow) crest travels to the right end (pictures A and B), where it is reflected as a crest (picture C). Remember that the pulse is not inverted at a free end and that the crest is higher at the end of the string, as discussed in connection with Figure 8.2 of this chapter. These phenomena are due to the superposition of the incident and reflected pulses, which are shown as a single dashed line (the two lie on top of one another). The reflected crest travels back to the left end (picture D), which is fixed, so the pulse is reflected as a trough (picture E). The incident crest and reflected trough at the left end superpose to give a node at this end. Now, we have the original pulse traveling to the right as a trough (picture F). (Note that if we had placed another crest on the string when the first one had made a single round trip, then the two would cancel each other out.) When the original pulse reaches the right end of the string a second time (picture G), the pulse is now an incident trough and is reflected as a trough. Finally, when this original pulse gets back to the left end a second time (picture I), it is reflected as a crest; a second crest now can be put on the string in phase with the original one. Thus, the pulse must complete two round trips of the string before a matching pulse is put on the string. This process takes twice as long as for the string with two fixed ends, so the period of the vibration is doubled and the frequency is halved.

$$T = \frac{D}{v} = \frac{4L}{v} \qquad (8.3)$$

$$f = \frac{1}{T} = \frac{v}{4L} \qquad (8.4)$$

The reason that only the odd harmonics occur on a string with one fixed end and one free end can be seen most easily by looking at why there is no second harmonic. When the left end of the string is jiggled at twice the fundamental frequency, there is just enough time between successive upward jiggles for the first crest to make one round trip on the string. But on its second trip, the original pulse moves to the right as a trough so that the first and second pulses cancel one another. This cancellation between pulses happens for every frequency that is an even integer multiple of the fundamental frequency.

As mentioned above, there is another way of determining the frequencies and wavelengths of the traveling waves that will produce standing waves on a given string. The first method discussed above seems to give a greater understanding of why only certain frequencies will be produced when a string is driven by bowing, plucking, or hitting. The second method is useful in at least two other ways. First, it is probably easier to remember, and second, it is easier to extend to other situations. In this second method, we look first at the wavelength of the traveling waves that produces the standing wave, and then, we calculate the frequency of both the traveling and standing waves from the traveling wave's speed and wavelength.

When we look at a standing wave on a string, we see one or more *loops*. Each loop is produced by that part of the vibrating string that is between two nodal points. Thus, when the string vibrates in its first normal mode, there is only one loop. Each higher normal mode has one additional loop in its vibrating pat-

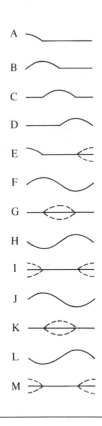

Figure 8.12. The pictures in Figure 8.10, but with pulses that have the correct width

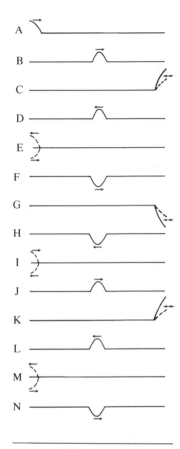

A

B

C

D

E

F

G

H

I

J

K

L

M

N

Figure 8.13. The movement of a single pulse on a string with one fixed end and one free end

tern. The patterns of strings with two fixed ends vibrating in each of the first four modes are displayed in Figure 8.14, and the loops show clearly. Since each loop is produced by either a crest or a trough of the traveling wave, within one wavelength there will be two loops. Another way of saying this statement is that the wavelength of the traveling wave is twice the distance between adjacent nodes. This situation is illustrated in Figure 8.15, where a single wavelength of the traveling wave is highlighted on the loops of the standing wave.

With this information, it is very easy to determine the wavelength of the traveling wave from the standing wave pattern. For example, Figure 8.16 shows a string that is 10 meters long and is vibrating in its fifth normal mode. You can see it is the fifth normal mode because there are five loops in the standing wave pattern. Dividing the 10-meter length by the number of loops tells us that the distance between adjacent nodes is two meters. Therefore, the wavelength is four meters (the total distance encompassed by two loops).

It also becomes quite easy to determine the frequencies and wavelengths of the normal modes of a string. When the string has two fixed ends, the wavelength of the first normal mode is twice the length of the string. The wavelength of the second normal mode is equal to the length of the string, but we can write this relationship as twice the length divided by two: $2L/2$. (We use this equation to emphasize the pattern that will emerge.) The wavelength of the third normal mode is $2L/3$ because the length of each loop is 1/3 the length of the string, and the wavelength is double the length of a loop. Now, we can look at the emerging pattern: $\lambda_1 = 2L/1$, $\lambda_2 = 2L/2$, $\lambda_3 = 2L/3$. This correspondence suggests correctly that the wavelength of the N-th normal mode is $2L/N$. To calculate the frequency of a particular normal mode, we use the information about the traveling wave's speed and wavelength and the relation that, by now, is familiar: The frequency is equal to the speed of the traveling wave divided by its wavelength. Thus, for example, the frequency of the fundamental f_1 is: $f_1 = v/\lambda_1 = v/2L$. Of course, we found this same answer using the first way of thinking about this situation. (See Equation 8.2.)

Now, let's go through the same reasoning for a string that has one fixed end and one free end. First, we note that at a fixed end, there is always a node, while there is always an antinode at a free end. Recognizing this behavior at the ends makes it easy to construct the normal mode vibration patterns for a string with ends of any type. Figure 8.17 shows the first five normal modes of vibration for a string with one fixed end and one free end. We can see that the first normal mode of vibration has only one-half a loop, so the wavelength of the traveling wave is four times the length of the string. (The wavelength is still the distance needed to encompass two complete loops.) Thus, the fundamental frequency of vibration will be equal to the speed of the wave on the string divided by four times the length of the string. (Compare this result to that given in Equation 8.4; they are the same, of course!) Note that if the string with two fixed ends and this string have the same wave speeds and same lengths, then the fundamental frequency of the one with two fixed ends will be double the fundamental frequency of this string. Saying this statement another way, the period of vibration for this string would be twice as long as the period of vibration for the string with two fixed ends. We saw this effect during the discussion of the first way of describing standing waves. For the string with different ends (one fixed and one free), a pulse must make *two* round trips on the string before being matched with another pulse. Only *one* round trip is required for a pulse that travels on a string with two ends the same, however. Thus, if the wave speeds and string lengths are the same, then the period of vibration for the string with different ends is twice that for the string with identical ends.

By inspecting the normal modes pictured in Figure 8.17, we can see that there are only odd harmonics. As mentioned above, the fundamental has one-half a loop. The second normal mode has 1½ loops (or three times the number of loops in the fundamental). Consequently, the wavelength is one-third as large and the frequency is three times the fundamental's values. Similar reasoning applies for each of the higher normal modes.

It now should be clear why the pitch of the sound produced by a vibrating string can be adjusted by changing the string's tension. When the tension is increased, the wave on the string goes faster, so the time for a round trip decreases. Thus, the period of the vibration is less, and the frequency rises. The fractional change is the same for all normal modes, so if they started out as harmonics, they remain harmonics. As noted in Chapter 6, the speed of a wave on a string is proportional to the square root of the tension. Thus, increasing the tension by a factor of four will double the wave speed, which, in turn, will double the frequency so the pitch will go up by one octave.

Organ Pipes

There are other systems besides strings for which we can build standing waves out of two traveling waves moving in opposite

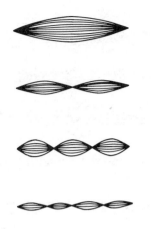

Figure 8.14. The first four normal modes of a vibrating string with two fixed ends

Figure 8.15. A string vibrating in its fifth normal mode with a single cycle of a traveling wave highlighted

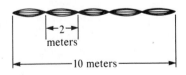

Figure 8.16. A string 10 meters long vibrating in its fifth normal mode

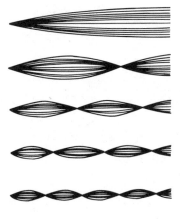

Figure 8.17. The first five normal modes of a string with one free end and one fixed end

directions. A sound wave in a straight pipe is the one system most similar to the string we just studied. This type of device is the fundamental component of pipe organs, which often are found in large churches and cathedrals, so we call it an organ pipe.

All organ pipes have a sharp lip at the side near one end. When air is blown against this lip, pressure vibrations are set up in the air inside the pipe. Thus, blowing air against this lip is analogous to bowing a string. Also, just as with a string, an organ pipe can have two extreme types of ends: open ends and closed ends. However, there is always an opening where the lip is located, so we have two types of organ pipes—open and closed—as determined by the pipe's other end.

Figure 8.18 shows diagrams of each of these pipes. It is important to recognize is that in an open pipe, both ends are open; this setup is like a string having two fixed ends (in each case, the ends are the same). A closed pipe, of course, has different types of ends and, thus, is similar to a string with one fixed end and one free end.

To develop our knowledge of standing sound wave patterns in an organ pipe, we will use the "loop method" that we used for strings. First, we need to look at which type of end has nodes and which has antinodes. There is, however, a slight complication: We can describe sound waves either in terms of the displacement of the medium or in terms of pressure variations. The complication arises because pressure nodes occur where displacement antinodes are located, and vice versa. It is easy to see why this

Figure 8.18. Organ pipes

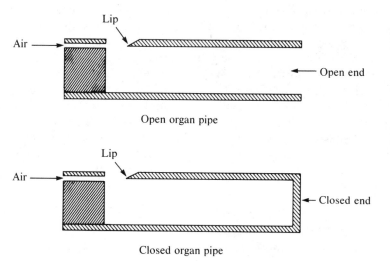

Open organ pipe

Closed organ pipe

An organ pipe that is used in the lab. Note the sharp lip that is struck by air blown in the mouthpiece.

statement is true if we look at what happens to the medium at the end of the pipe. Near an open end, for example, the air can move easily in and out of the pipe. Thus, displacement antinodes occur at open ends of pipes. However, there is no way the pressure can vary at an open end. If the pressure starts to increase, air will just be pushed out of the pipe, and air will rush in if the pressure drops. Thus, we have a pressure node (and a displacement antinode) at an open end of a pipe. Just the opposite effects occur at a pipe's closed end: There is a pressure antinode and a displacement node. We find these conditions are always true for standing sound waves; that is, pressure nodes and displacement antinodes always occur at the same location, and pressure antinodes always are located where there are displacement nodes.

As we have discussed in Chapter 6, sound waves are longitudinal. Thus, the movement of the medium in an organ pipe occurs along the length of the pipe. However, it is convenient to show the standing wave patterns in organ pipes with diagrams that look similar to those drawn for strings. Such a diagram is shown in Figure 8.19 for the second normal mode in an open pipe. In this diagram, the vertical distance between the two curved lines at any point shows the relative vibration amplitude of the standing wave at that point. The two locations where the curved lines cross are displacement nodes; these locations are marked with an N above the pipe. The three locations marked A are displacement antinodes. As discussed above, pressure nodes occur at the points marked A, and pressure antinodes exist at the two displacement nodes. Even though the distance between the two lines is perpendicular to the pipe, we must remember that the actual vibration of the air is directed along the pipe. Note that in Figure 8.19, displacement antinodes are shown at each open end of the pipe.

In Figure 8.20, the loop diagrams for the first four normal modes in both an open pipe and a closed pipe are shown. Once a

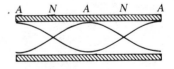

Figure 8.19. A "loop" diagram showing the standing wave pattern for the second normal mode in an open pipe

Figure 8.20. Loop diagrams for organ pipes

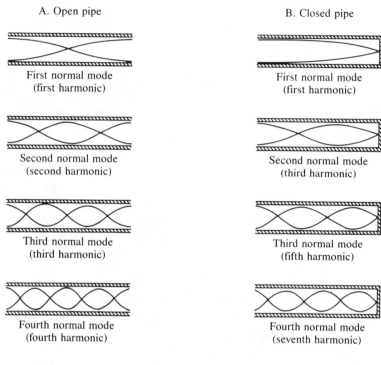

A. Open pipe

First normal mode
(first harmonic)

Second normal mode
(second harmonic)

Third normal mode
(third harmonic)

Fourth normal mode
(fourth harmonic)

B. Closed pipe

First normal mode
(first harmonic)

Second normal mode
(third harmonic)

Third normal mode
(fifth harmonic)

Fourth normal mode
(seventh harmonic)

loop diagram for a standing sound wave has been constructed, we proceed to determine the wavelength and frequency of the wave in the same way we did for strings. That is, the wavelength of the traveling waves (which produce the standing wave) is twice the length of one loop, and the frequency is the speed divided by the wavelength. Thus, we obtain the same relations between closed and open pipes as we did with the two types of strings. These relations are summarized in Table 9. A particular point of interest in this table is that the fundamental frequency of a pipe or string with different ends is half the fundamental frequency of the pipe or string with ends alike. Thus, the note sounded by a closed organ pipe is one octave below that of an open pipe having the same length. A second general result concerns the vibration recipes of these systems. When the pipe or string has the same type of ends, then all harmonics can be created. However, when the string or pipe has different ends, only the odd harmonics can be produced.

In this chapter, we have discussed at great length why only certain frequencies will produce normal mode vibrations for strings. We then used the same type of reasoning for understanding the normal modes of standing sound waves in pipes. We

Table 9. Similarities of standing wave patterns

Description of Vibrating System	Wavelength of Nth Normal Mode	Frequency of Fundamental	Possible Vibration Recipes
Vibrating string with two free ends	$\dfrac{2L}{N}$	$\dfrac{v}{2L}$	All harmonics
Vibrating string with one free and one fixed end	$\dfrac{4L}{2N-1}$	$\dfrac{v}{4L}$	Only odd harmonics
Open organ pipe (two open ends)	$\dfrac{2L}{N}$	$\dfrac{v}{2L}$	All harmonics
Closed organ pipe (one open and one closed end)	$\dfrac{4L}{2N-1}$	$\dfrac{v}{4L}$	Only odd harmonics

should recognize that the normal modes of even more complicated systems such as Chladni plates also can be understood in terms of standing waves. Just as for strings, the standing waves are formed from the superposition of traveling waves moving in opposite directions. Several other examples of standing waves will be discussed in the next chapter.

Summary

Whenever two waves of the same frequency and amplitude are traveling in opposite directions, they superpose to produce a standing wave. If the amplitudes differ, then we get only a partial standing wave. One convenient way of producing two identical waves traveling in opposite directions is with an incident and a reflected wave. The amplitude and phase of the reflected wave compared to the original or incident wave depends on the boundary. We saw various cases of this statement for the end of a string; the two extreme cases were a fixed end (for which there is a 180° phase change) and a free end (which produces no phase change).

The normal mode vibration patterns of extended systems can be explained using the concept of standing waves. Two particularly easy systems to study (because the traveling waves are constrained to move in only one dimension) are a string and an organ pipe. In any such system, traveling waves that are reflected back and forth from each end superpose to produce the normal modes as standing waves. The types of ends determine the types of standing waves produced: Ends that are the same can produce any harmonic while different ends can produce only odd harmonics. The specific frequencies of the normal modes are determined by the length of the system (string or pipe) and the speed of the traveling wave as well as the types of ends.

An easy way to keep track of standing wave patterns and to relate them to the wavelengths of traveling waves is through the use of loops. A loop extends from one node to the next, and the traveling wave wavelength is twice the length of a loop. The allowable loop patterns are easy to construct for various types and combinations of ends. At a string's fixed end and a pipe's closed end, there are displacement nodes, while antinodes occur at the free end of a string or the open end of a pipe.

Suggested Readings

Organ pipes are discussed in greater detail than our discussion in the article, "The Physics of Organ Pipes," *Scientific American,* January 1983. The way the air vibrates as it passes the lip of the pipe is detailed in this article. In addition, the vibration recipes that occur for different shapes of organ pipes are discussed.

Review Questions

Figure 8.21. For use with review question 1

1. Figure 8.21 shows two specially shaped pulses moving through each other on a string. The solid line indicates the string; it is missing in the region of the dashed lines that show the pulse shapes. Which of the following shows the correct string shape at this instant of time?

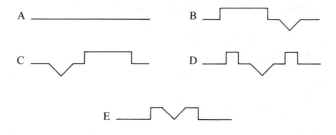

    a. A

    b. B

    c. C

    d. D

e.   E

2.   In Figure 8.22, a square pulse is sent along a string toward a free end. Which of the following does the reflected pulse look like?

Figure 8.22.   For use with review question 2

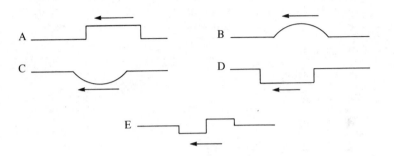

a.   A

b.   B

c.   C

d.   D

e.   E

f.   A square pulse cannot be reflected from a free end.

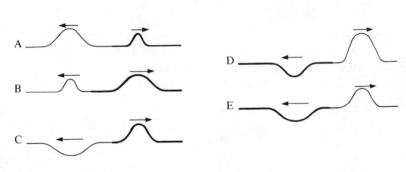

Figure 8.23.   For use with review question 3

3.   Which of the diagrams in Figure 8.23 shows correctly the reflected and transmitted pulses that are the result of an incident pulse striking a junction between two strings from the left?

a.   A

b.   B

c.   C

d. D

e. E

4. A wave with a wavelength of 4 meters travels along a string at a speed of 36 meters per second and reflects off a fixed end. How far from the end of the string is the first node (not counting the one at the end) of the standing wave?

a. 1 meter

b. 2 meters

c. 4 meters

d. 8 meters

e. 9 meters

5. A wave with a frequency of 30 Hz travels along a string at a speed of 36 meters per second and reflects off a free end. How far is the first node from the end of the string?

a. 0.2 meters

b. 0.3 meters

c. 0.4 meters

d. 0.6 meters

e. 1.2 meters

6. An 80 Hz wave is sent along a string toward a fixed end. When the wave reflects, a node forms at a distance of 10 cm (0.1 meters) from the end of the string. What are the speeds of the incident and reflected waves?

a. 4 meters per second

b. 8 meters per second

c. 16 meters per second

d. 32 meters per second

e. 64 meters per second

7. A space traveler lands on a distant planet and wants to determine the speed of sound in the planet's atmosphere. She sets up a loudspeaker 100 meters from the base of a cliff and plays a 150 Hz tone. She walks from the base of the cliff toward the loudspeaker and observes loud sounds at the cliff and at a point 25 meters from the cliff. What is the speed of sound?

a. 2,500 meters per second

b. 5,000 meters per second

c. 7,500 meters per second

d. 15,000 meters per second

e. 30,000 meters per second

8. A string is 0.6 meters long, has two fixed ends, and vibrates in its third normal mode. What is the wavelength of the waves traveling along this string?

a. 0.2 meters

b. 0.3 meters

c. 0.4 meters

d. 0.6 meters

e. 1.2 meters

9. A nine-meter-long rope hangs vertically from a window. It takes 3 sec for a pulse to travel from one end of the rope to the other. With what period should we jiggle (with small amplitude) the top end of the rope in order to have the rope vibrate in its second normal mode?

a. 1 sec

b. 2 sec

c. 3 sec

d. 4 sec

e. 6 sec

10. A string with two fixed ends vibrates at 880 Hz in its fourth normal mode. The string is 0.6 meters long. How fast do waves travel on this string?

a. 132 meters per second

b. 264 meters per second

c. 396 meters per second

d. 528 meters per second

e. 1,056 meters per second

9

Examples of Standing Waves

Important Concepts

- A seich is a standing water wave. The standing wave patterns are determined by the ends of the water container.

- The musical sound produced by an instrument depends on standing waves. They occur on a string or in a column of air within the bore of a wind instrument.

- The mechanism (bow, reed, or lips) that drives the basic vibrating object (string or air column) produces many different frequencies. Only the resonant frequencies of the vibrating string or air column build up and are sustained.

- Different musical notes are obtained either by changing the length of the vibrating object (the string or air column) or by exciting only higher harmonics.

- The vibration recipe of a violin is influenced strongly by the resonance box and bridge.

- The shape of the bore for a wind instrument strongly influences the vibration recipe of the sound produced.

173

We have discussed in detail how two waves traveling in opposite directions along a string will create standing waves on that string. We also have looked at standing wave production in a straight pipe, and we have stressed that standing waves always will be formed whenever two waves of the same frequency and wavelength travel in opposite directions. Now, we wish to look at some additional examples of this process of creating standing waves from the superposition of traveling waves. As a first example, we will investigate seiches, which are standing waves on enclosed bodies of water such as bays and harbors. Then, we will turn our attention to musical instruments. We will discuss various stringed instruments, woodwinds, and horns. I must emphasize that the concepts introduced in the description of standing waves on the string are still valid in all these cases. There, of course, will be some differences in details because of the different types of waves involved.

Seiches in an aquarium. The first normal mode is shown in A and B, and the second normal mode is shown in C and D. The water has been colored with ink to make it more visible.

Seiches

When a standing water wave occurs in a harbor or bay, the wave is called a *seich* (pronounced sāsh). In the laboratory, we easily

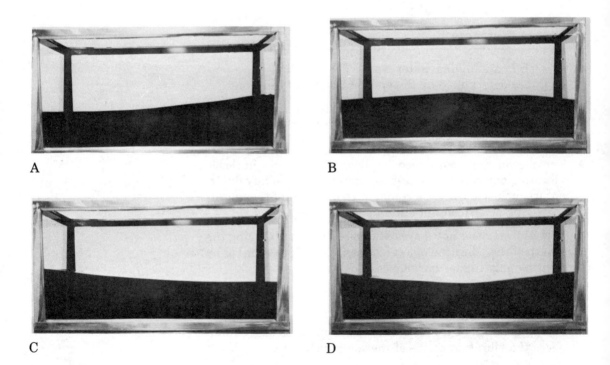

A

B

C

D

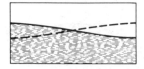
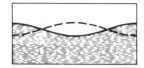

A. First normal mode   B. Second normal mode

Figure 9.1. Normal modes of seiches in an aquarium

can produce a seich in an aquarium. With 4–6 inches of water in the aquarium, we use a sponge (or a flat piece of wood) to generate a periodic disturbance on a small part of the water's surface. Alternately pushing down and lifting up on the sponge causes the water level to rise and fall. Just as with any standing wave, when the frequency of this periodic disturbance matches a natural frequency of the system, resonance will occur, and a standing wave will develop. The only questions that remain concern the details of this process. We must look at how water waves reflect off the aquarium edges, what determines the speed of the waves moving back and forth within the system, and how we can transfer this knowledge to what happens in real bays and harbors. First, however, we will look at the normal modes of vibration for these seiches in the aquarium.

In Figure 9.1, we see the vibration pattern of the water as it oscillates in its first and second normal modes. These pictures look just like the patterns we had for the first two modes in an open organ pipe. (Here, however, the water's surface actually takes the shapes shown by the solid and dashed lines—at opposite points in the cycle, of course.) It is clear from these pictures that a displacement antinode is located at each end of the aquarium. The actual movement of the water during each vibration cycle cannot be deduced from the pictures in Figure 9.1. Some simple experimenting with the seich in our aquarium shows that the water near the surface does not move horizontally very much. The major portion of the water movement occurs below the surface; there are large currents under each node as the water moves from one antinode to the other and back again during each cycle. This movement is indicated in Figure 9.2.

As we have stressed, all standing waves are produced from the superposition of two oppositely directed traveling waves. Thus, we are interested in seeing how our knowledge about the speed of the moving water waves can be applied in this case. For the first normal mode standing wave, the length of the aquarium equals the length of one loop (actually, two half loops). Thus, the wavelength is twice the length of the tank, or about 40 inches for a typical aquarium. This relationship means (for a 4–6 inch

Figure 9.2. The movement of water in a seich

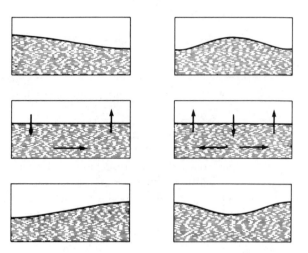

depth) that the depth-to-wavelength ratio falls into the shallow water wave category, and deeper water will produce faster waves. We, therefore, can do a simple experiment to show the effect of the wave speed on normal mode frequency. Recalling this dependence for the first mode as given in Equation 8.2, $f = v/2L$, we expect to find the frequency decreasing as we make the water shallower. (This experiment is very similar to one in which we observe the change in the fundamental frequency of a string when we decrease the string's tension.)

For seiches in harbors or bays, the periods of the normal modes are usually quite long, on the order of 3–5 minutes for the fundamental. Consequently, the effects are difficult to observe. Willard Bascom used stop-action photography to accentuate the effects. By exposing one frame of a movie film every second and projecting it at the normal rate, he speeded up the action by a factor of 16. He used this technique to photograph Monteray harbor from a bluff overlooking the harbor. When the film was shown, it was easy to see the boats at anchor in the harbor swinging rhythmically back and forth as the water executed its standing wave motion. This movement was happening, of course, even without the stop-action technique; the movement just was not noticeable because the changes in boat positions occurred so slowly.

As noted previously, a seich has an antinode at a vertical wall of the container holding the water, whether the container is an aquarium or a bay. Other types of container edges occur for bays, however, so we need to discuss the effect of these various boundaries on seiches. A side of a bay open to the ocean, such as the one indicated by the dashed line in Figure 9.3A, is just the opposite of a vertical wall. The presence of the ocean forces a node along this edge of the bay. This situation is easy to understand by using an argument similar to the one we used when describing why pressure nodes must occur at open pipe ends. In this case of the harbor and the ocean, we note that if the water along the line where the harbor meets the ocean starts to rise, it will flow out into the ocean. On the other hand, the ocean will rush in to fill up any voids that develop. Thus, the water level at the ocean edge of the bay must remain even with the sea level, and we have a node. A third type of edge, for a bay, is a gently sloping beach. However, as discussed in Chapter 7, no waves are reflected from this type of edge; the energy in the incident waves is dissipated when the waves break. Since there are no reflected waves, standing waves (seiches) will not be formed.

Thus, we see a great similarity between seiches and standing waves on strings or in pipes. We have two types of edges in a bay (just as there are two types of ends for strings and pipes). So, the two patterns of standing waves—with edges (or ends) either alike or different—that we met in the last chapter can occur as seiches also. As an example, look at the bay off an ocean shown in Figure 9.3. From the map in Figure 9.3A, we can see that the bay has land on the north, east, and south sides. We cannot tell from this map if these edges are vertical walls or sandy beaches, but from the cross-sectional view of the ocean, bay, and land in Figure 9.3B, we know the east edge is a steep cliff. Assuming the north and south walls are also steep cliffs, we then have the

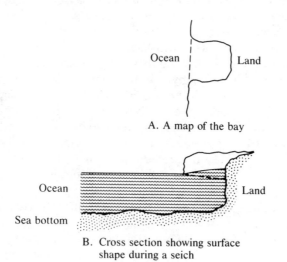

Ocean   Land

A. A map of the bay

Ocean   Land

Sea bottom

B. Cross section showing surface
shape during a seich

Figure 9.3. The first seiching mode for a bay adjoining the ocean

Figure 9.4. The violin

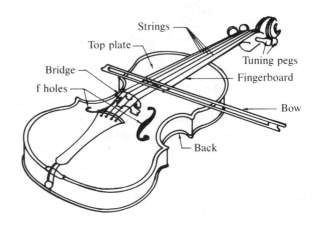

A violin. Note the f holes in the top of the violin; they show up well in this view. (Photograph provided courtesy of Robert Benedetto, violin maker, Clearwater, Fla.)

possibility that each type of seich occurs in this bay. In the north-south direction, vertical walls at each end means the seich's normal modes will be similar to those in an aquarium (or an open pipe). The pattern of seiching normal modes in the east-west direction will be like those of a closed pipe because of the different types of "ends." The fundamental mode of an east-west seich is illustrated in Figure 9.3B.

Stringed Instruments

We now will look at the design of various stringed instruments from a physics point of view, using the knowledge we have gained in the last several chapters. We discussed the piano in Chapter 4, and we noted that longer, heavier strings at lower tensions are used for the low notes. (It is also true, of course, that heavier, longer strings are used for low notes on all stringed instruments.) Now, we can see some reasons for this setup. The strings are longer so that the traveling waves producing the fundamental will take a longer time to move back and forth. The strings are heavier and have a lower tension so that the wave speed will be slower to also increase the transit time for the traveling waves. As we know, when a string is vibrated—either by bowing, hitting, or plucking—many frequencies are excited. However, only those frequencies for which standing waves are produced will be maintained and will produce a significant sound. All stringed instruments use strings that are fixed at each end. Thus, the vibration recipe of stringed instruments can contain all the harmonics.

As a second example of a stringed instrument, we consider

the violin. Just as with the piano, we will look at how the violin produces its unique sound. (Other members of the violin family, the viola, cello, and double bass, operate in a way that is similar to the violin. The size differences among these instruments give them different frequency ranges.) As we can see in Figure 9.4, the violin has four strings mounted on a resonance box. The box is constructed from carefully conditioned and shaped wood pieces. The top plate, with its f holes, and the back are particularly important parts of the resonance box. The strings can be driven either by bowing or plucking, but we only will talk about the bowing mode here. The strings are tuned to the notes G_3, D_4, A_4, and E_5. A player obtains other notes by using a finger to tightly clamp a string against the fingerboard. This clamping shortens the string so that the string's normal mode frequencies are raised. The strings, made of gut or very thin wire, are much more flexible than those on a piano, so the normal modes are harmonics. Just as we saw with the piano, a violin string vibrating by itself will be barely audible because it does not couple well to the air. To correct this deficiency, the string passes over and drives the bridge, which, in turn, drives the entire box. However, the violin box does not just amplify the sound produced by the string. Instead, the quality of a violin's sound is affected strongly by the resonance properties of the bridge, the top plate, and the back.

The neck (with fingerboard) being clamped onto the body of a violin during construction. (Photograph provided courtesy of Robert Benedetto, violin maker, Clearwater, Fla.)

The best violins ever made were those produced in Cremona, Italy, by such craftsmen as Andrea Amati, Antonio Stradivari, and Giuseppe Guarneri during the period of 1650–1750. These prized instruments are coveted by musicians for their excellent tonal quality and ease of playing. These violins have become collector's items and currently cost more than $100,000 each. For years, instrument makers and scientists have been trying to unlock the secrets of these magnificent instruments. However, the techniques used by those craftsmen were lost when they died. Recently, some significant progress has been made in understanding how the many parts of a violin work together to produce a high quality sound. These ideas serve as an excellent example of standing waves and resonance, so we will discuss the ideas in some detail.

When the string is bowed, it vibrates horizontally and drives the bridge. As a result, the bridge rocks sideways, and the violin box vibrates in three different ways for three different frequency ranges. To understand these different vibrational modes, we first must look at some construction features, which

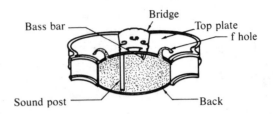

Figure 9.5. Constructional details of a violin

are illustrated in Figure 9.5. The bridge, which is held in position by the tension of the strings, has two feet that rest directly on the top plate of the resonance box. A wooden beam (the *bass bar*) is glued to the lower surface of the top plate directly under that bridge foot which is next to the lowest frequency (bass) string. Under the bridge's other foot, the *sound post* is wedged between the top plate and the back.

For low frequencies, the sideways motion of the bridge twists the top plate about a longitudinal axis, as shown in Figure 9.6. The vibrations also are transferred to the back through the sound post, causing the back to move up and down. Note that—just as we should expect—large pieces of the violin's resonance box respond to this low frequency oscillation. Vibrations having frequencies in an intermediate range cause the top plate to twist about a transverse axis as it vibrates. (See Figure 9.7.) In this mode, the bass bar provides the necessary rigidity so that the top plate will rock rather than bend. Also, the sound post now remains stationary because the large back plate cannot move rapidly enough to follow these vibrations. Finally, for high frequencies, only that section of the top plate that lies between the f holes vibrates as shown in Figure 9.8.

When any of these different sections of the resonance box vibrates, it drives the air inside. Thus, the volume and shape of the air inside the box and the size and shape of the f holes will add new resonances to the entire vibrating system of the violin. The bridge (See Figure 9.9) also will affect the violin's sound. For purposes of analysis, we can interpret the bridge's shape as a complicated set of springs and weights that have many resonant frequencies. By slightly varying the thickness of the bridge at different points, we can alter its response at these frequencies and significantly affect the tone quality produced by the violin.

We have seen how complicated the structure of the violin is and how all its components can affect one another. The violin maker must adjust the various parts of the instrument during its construction so that these resonances do not get out of control.

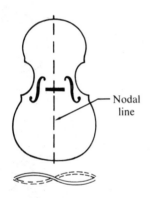

Figure 9.6. Vibrational mode of top plate for low-range frequencies

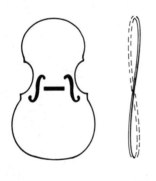

Figure 9.7. Vibrational mode of top plate for mid-range frequencies

The types of adjustments are similar to ones we might do on a simple spring-mass system to change its response. On this much simpler system, we could change either the mass of the vibrating object or the spring stiffness to alter the natural frequency, or we could change the damping of the system by modifying the size and shape of the mass or spring. We also can change the response of this simple system by varying the coupling between it and whatever is driving it. As an example of how a craftsman adjusts the natural frequencies of the back of a violin, I show an enlarged cross-sectional view of a violin's back in Figure 9.10. By removing wood in the middle of the back—in the region around the point marked A—the mass will be reduced, causing a higher resonant frequency. On the other hand, the wood in the region around B acts as a spring, and any reduction in its thickness will reduce its stiffness, decreasing the natural frequency. Both of these changes also can affect the damping of various normal modes, however, so the adjustments are critical. How well these types of adjustments are made determines the quality of the finished instrument.

When playing a violin, the performer moves the bow across the string in either an upward or a downward direction (called *up bow* or *down bow* movement, respectively). As the bow moves, energy is being supplied continuously to the vibrating string, and the string seems to be taking the shape of the fundamental as shown in Figure 9.11. However, the string actually is vibrating in many harmonics at the same time. High speed photography of a bowed string has revealed what the eye cannot see: The string is bent into two straight pieces at all times. This fact first was discovered by Herman von Helmholtz in 1860. He used some clever experimental techniques, which included mechanically generating Lissajous figures.

The junction of the two straight pieces of string races around the edge of the apparent fundamental mode shape once during every period of the fundamental. This activity is illustrated by the movement of point P in the sequence of pictures of Figure 9.12. The bow transfers energy into the string by alternately grabbing and releasing the string. The vibrations of the string, itself, control when the string sticks to the bow (and is pulled along by the bow) and when the string slides back (against the

Figure 9.8. Vibrational mode of top plate for high-range frequencies

Front view   Side view

Figure 9.9. Detail of a violin bridge

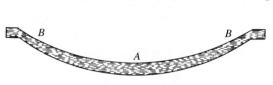

Figure 9.10. Cross section of a violin back

Figure 9.11. The appearance of a bowed string

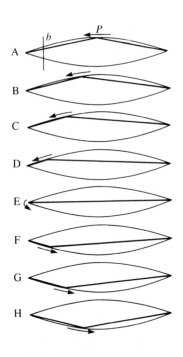

movement of the bow). Whenever the point P moving around its path (as pictured in Figure 9.12) crosses the place on the string where the bow is located, the condition of sticking or sliding reverses. As an example, the bow in Figure 9.12 is moving across the string at the place marked b, which is located ⅛ of the way from the string's left end. Point P then coincides with this location in pictures D and F. Thus, the string is being pulled along by the bow in pictures A–C. Then, at picture D, the string changes; it starts sliding backwards. It continues sliding until picture F, where the bow again grabs the string. There is more time between picture F and the picture analogous to D in the next cycle than there is between pictures D and F as shown here. Thus, the string slides backwards more rapidly than when it is being pulled along. The important aspect of this effect from our point of view is the control exercised by the string's vibration on the driver (the bow). We will see this same type of effect in wind instruments.

In the terms we used in Chapter 2, the string and the bow are strongly coupled. The bow drives the string, but in this case, the bow does not have a unique driving frequency. Instead, the bow causes the string to vibrate over a large range of frequencies at the same time. The string then resonates at only its normal mode frequencies. Certain harmonics of the string resonate more strongly than others because of the control exercised over them by the resonance box (the body of the violin). This complicated interconnection of drivers and oscillating systems makes the design and construction of an excellent violin extremely elusive.

While scientists understand the fundamental aspects of producing sound with a violin, the actual production of beautiful music with the instrument remains a combination of several arts. The virtuoso performer who coaxes the ultimate out of any instrument complements the craftsman who produces a violin with exceptional quality. As with any artistic expression, there is a subjective appraisal by the audience. It is extremely difficult for science to quantify this subjective evaluation. However, in principle, it is possible to measure the sound produced and to compare the frequencies and vibration recipes of various instruments.

Figure 9.12. The actual shape of a bowed string. Adjacent pictures are separated in time by 1/16 of a period. Bowing occurs at point b on the string. (Adapted from "The Physics of the Bowed String" by John C. Schelleng. Copyright © 1974 by Scientific American, Inc. All rights reserved.)

Wind Instruments

The woodwinds and brass instruments are similar to organ pipes in that the basic vibrating element is the column of air inside the instrument. The part of the instrument that contains the vibrating air column is called the *bore*.

There is a fundamental difference between the organ pipes we discussed and other wind instruments, however. We looked at organ pipes that have cylindrical bores. That is, they have the same cross-sectional size all the way along the pipe. In contrast, the bore of most instruments changes diameter along its length. (This changing in diameter also happens with some organ pipes, but we will not be discussing them.) A cone is another simple bore shape, and for this bore shape, the diameter changes constantly along the length of the bore. Many instruments have bores that combine conical and cylindrical sections, and some are even more complicated. The bore shape helps to determine the vibration recipe of the tones produced. We saw that an open organ pipe can have all harmonics in its vibration recipe while closed pipes can have only odd harmonics. The vibration recipe for a conical bore that is closed at its small end also contains all the harmonics. Thus, different combinations of conical and closed cylindrical bores (that is, different lengths of each type) will produce different vibration recipes. Just as with strings, a longer vibrating system (that is, a longer bore) produces a lower pitch. The new elements of sound that must be considered for wind instruments are the driving mechanism, the way the air column length is controlled, and the vibration recipe. Since these are fundamentally different for woodwinds and brass instruments, we will consider them separately.

Woodwinds

For woodwinds, there are two ways for a player to produce vibrations of the enclosed air column, depending on the instrument. Oboes, clarinets, saxophones, and bassoons are reed instruments, while air is blown across an opening to play a flute or a piccolo. In both of these situations, a continuous range of frequencies is produced by blowing, but only those frequencies that resonate with the rest of the instrument will build up and be sustained. For the reed instruments, the vibrations in the bore are produced by the movement of the reed. Once started, the frequency of the vibrating reed is controlled by the standing wave in the instrument. Blowing across the opening of a flute or

An oboe. Note the small mouthpiece that has a double-sided wooden reed. (Photograph provided courtesy of The Selmer Company, Elkhart, IN, U.S.A.)

piccolo produces vibrations in a way similar to the method of producing vibrations in an organ pipe. That is, blown air hitting the edge of the hole either goes into the instrument or outside, creating alternate raised and lowered pressures. Just as with other instruments, the vibrating column of air in the bore regulates the rate of these pressure variations.

The various woodwinds have different sizes, so their frequency ranges are different. These differences are indicated in Figure 9.13. In each of the woodwinds, different notes are played by changing the length of the air column through the use of finger holes. Both pads actuated by keys and actual fingers are used to open or close holes along the side of the instrument. Through trial and error, instrument makers have determined the locations and sizes of the holes needed to create desired pitches. Long after the development of these instruments, scien-

Figure 9.13. Frequency ranges of various instruments

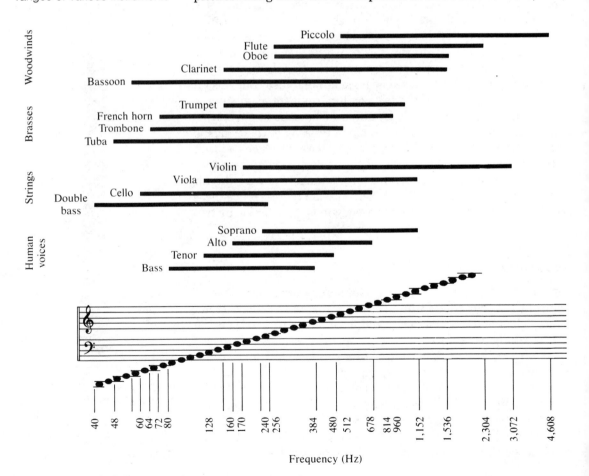

Frequency (Hz)

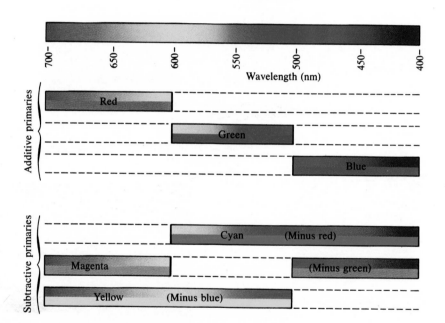

Plate 3. Examples of color addition

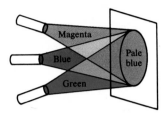

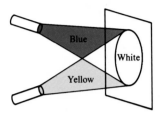

A. Adding magenta, blue, and green

B. Adding blue and yellow

Plate 4. Examples of color subtraction

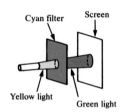

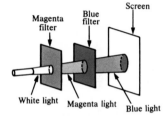

A. Yellow light through a cyan filter

B. White light through a magenta and a blue filter

Plate 5. A painting of the chromaticity diagram of Figure 12.4 (Courtesy of Eastman Kodak Company.)

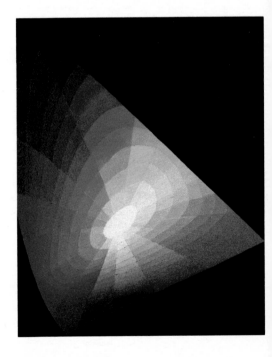

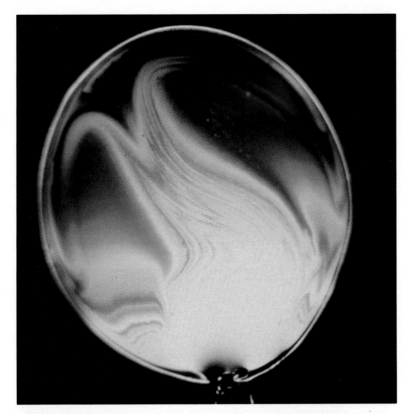

Plate 6. A thin soap film showing interference effects in the reflected light. The film at the top is so thin that destructive interference is occurring for all colors.

Plate 7. The colorful effects produced by stressed plastic between crossed Polaroid filters (The plastic is optically active and rotates the plane of polarization varying amounts according to wavelength and the stress of the plastic.)

Plate 8. A spectacular scene in nature. The more pronounced primary rainbow is accompanied by a secondary bow on the outside and several faint supernumerary arcs just inside the primary bow. (Photograph courtesy of Alistair B. Fraser.)

Plate 9. A corona around the moon (From *Rainbows, Halos, and Glories* by Robert Greenler, Cambridge University Press, 1980. Photograph by Robert Greenler.)

Plate 10. A halo around the sun. The tree in the foreground covers the direct light from the sun so the film will not be overexposed. (From *Rainbows, Halos, and Glories* by Robert Greenler, Cambridge University Press, 1980. Photograph by Robert Greenler.)

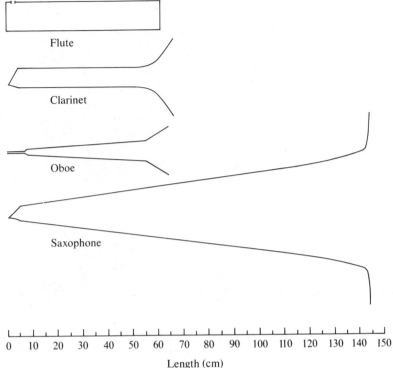

Figure 9.14. The shapes of some woodwinds (The diameter is multiplied by a factor of five compared to the length.) Illustrations based on figures 42, 49–51 (from *Horns, Strings, and Harmony* by Arthur H. Benade. Copyright © 1960 by Educational Services Incorporated. Reprinted by permission of Doubleday & Company, Inc.)

A clarinet. The mouthpiece has a wide wooden reed mounted on the back side. (Photograph provided courtesy of C.G. Conn, Ltd., Elkhart, IN.)

tists have studied them and have discovered some interesting facts. For example, the length of the vibrating air column does not stop at an open hole. Instead the column extends slightly farther; the actual extra length depends on the size of the opening compared to the size of the instrument bore at that point. As we saw earlier, the vibration recipe of the instrument is influenced strongly by the shape of the bore.

Some examples of woodwind shapes are shown in Figure 9.14. Notice that the oboe and saxophone have approximately conical bores (with a bell added at the open end). The normal modes that can exist in a conical air column are harmonics, and, as mentioned earlier, all of them are present. The bell serves as a coupler between the vibrations inside the instrument and the outside air, particularly for the low notes. The vibrations of higher notes are passed to the outside air through the holes, which are opened to shorten the vibrating air column. The flute, on the other hand, has a cylindrical bore that is open at each end. Thus, the flute acts like an open organ pipe and has the possibility for all harmonics to be present in the vibration recipes of

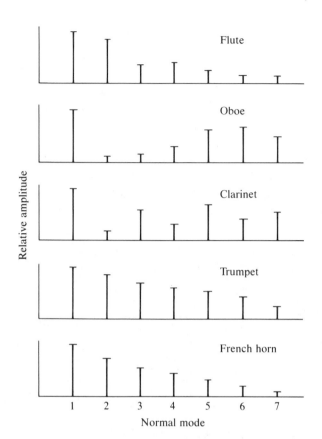

Figure 9.15. Some typical vibration recipes for various musical instruments

A saxophone. The mouthpiece and reed are very similar in construction to those of the clarinet. (Photograph provided courtesy of The Selmer Company, Elkhart, IN, U.S.A.)

its notes. It does not need a bell because the holes are so large compared to the diameter of the bore that the vibrations are transmitted easily to the outside air. Finally, the clarinet has a cylindrical shape with the reed at one end. Thus, the clarinet performs very much like a closed organ pipe. The closed end is at the reed, so the even harmonics are suppressed in the vibration recipes of notes played with a clarinet.

We have mentioned how the shape of the bore influences the vibration recipe and, therefore, the tone quality of an instrument. There are other factors that also affect this aspect of sound, such as how hard the player blows, how tightly he squeezes the reed between his lips, and which frequency is being played. Soft tones on any instrument always have fewer and less pronounced overtones in the vibration recipe. Essentially, the fundamental mode is the only one excited for extremely soft tones. Also, notes in the lower part of an instrument's range usually have a vibration recipe that is closer to that expected theoretically than do notes in the higher end of an instrument's frequency range. Some typical vibration recipes are shown in Figure 9.15. These have been chosen because they show some indication of the predicted patterns. We can see a suppression of the even harmonics for the clarinet in comparison to the other instruments shown, while the oboe has several overtones with larger amplitudes than the fundamental. In both cases, the distinctive quality of the resulting sound is determined by the respective pattern of the vibration recipe.

The vibrations in the air column control the vibrating of the reed or the air blowing across the open hole. This relationship gives an effect that is similar to the one we saw above with the violin where the vibrating string controlled how the bow grabbed and released it. In each case, the length of the vibrating element determines the wavelengths of the standing waves. The frequencies of these standing waves are, in turn, fixed by these wavelengths and the speeds of the traveling waves. Finally, in each case, there is a feedback effect so that the driving mechanism vibrates at only the desired frequency.

A flute. To produce vibrations of the air column inside the flute, air is blown across the opening at the left end. (Photograph provided courtesy of C.G. Conn, Ltd., Elkhart, IN.)

Brass Instruments

The brass instruments in an orchestra include the French horn, the trombone, the trumpet, and the tuba. In each of these instruments, the length of the bore (and, therefore, the length of the vibrating air column) can be changed with valves or a slide. The bugle is a simpler brass instrument, which has no mechanism for changing its length. Thus, one naively might expect that the bugle can play only one note. This, of course, is nonsense. A one-note instrument would generate very little interest within the musical community! Because of the bugle's simplicity, we will discuss the characteristics of the bugle and then indicate the changes introduced when a set of valves or a slide is used.

A bugle has an approximately conical bore with a flared bell at the open end and a mouthpiece at the closed end. Thus, all harmonics are present in the vibration recipes. The bugler can play different notes on this instrument of fixed length by changing how tightly he squeezes his lips together. To excite the higher harmonics (actually, to suppress the lower ones) so that higher pitched notes are sounded, a bugler squeezes his lips together more tightly. The length of the bugle is set so that the fundamental mode of vibration will produce the note C_3 with a frequency of about 131 Hz. However, this note (called the *pedal note*) is never used in bugle music. Instead, the notes of the sec-

A bugle. Note that the bugle has no valves or other moving parts. (Photograph provided courtesy of Buglecraft, Inc., Long Island City, NY.)

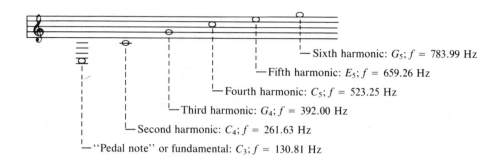

└─Sixth harmonic: G_5; $f = 783.99$ Hz

└─Fifth harmonic: E_5; $f = 659.26$ Hz

└─Fourth harmonic: C_5; $f = 523.25$ Hz

└─Third harmonic: G_4; $f = 392.00$ Hz

└─Second harmonic: C_4; $f = 261.63$ Hz

└─''Pedal note'' or fundamental: C_3; $f = 130.81$ Hz

Figure 9.16.   The notes of a bugle

A trumpet. The valves are used to change the length of the bore. (Photograph provided courtesy of C.G. Conn, Ltd., Elkhart, IN.)

A trombone. The loop on the right can be slid outward to increase the length of the bore. (Photograph provided courtesy of C.G. Conn, Ltd., Elkhart, IN.)

ond through sixth harmonics, as shown in Figure 9.16, are used. As can be seen, the notes that are used all fall within a two-octave range, while the unused fundamental lies a full octave below the lowest note that is used.

As noted above, the lips determine the vibrational mode of the instrument. Thus, the lips serve the same purpose in the brass instruments that the reeds do in the woodwinds. However, lips are heavier than reeds, so the vibrating air column does not exert quite as much control over the lips as it does over a reed in a woodwind. Therefore, a player can change slightly the vibrational frequency away from that determined by the length of the vibrating air column. However, there are limits to this frequency shift; it is less than that needed to move to the next higher or lower note. Consequently, only the notes displayed in Figure 9.16 can be played on a bugle. This deficiency is corrected in the orchestral brass instruments by using valves or a slide to change the length of the vibrating air column. In this way, the correct length for each note can be selected by the player, and the full set of twelve notes within each octave of the instrument's range can be played. The range of fundamental frequencies for several brass instruments also is shown in Figure 9.13.

There are other differences between the bugle and the orchestral brass instruments. One of the major ones is the shape of the bore. Whereas the bugle's bore was purely conical, the other instruments have a more complicated shape that is partly conical and partly cylindrical. The relative sizes and shapes of the trumpet, trombone, and French horn bores are shown in Figure 9.17. These different shapes and the different proportions of conical and cylindrical parts of the bores give each instrument its distinctive tonal quality.

Sample vibration recipes for the trumpet and French horn are given also in Figure 9.15. These show the influence of the conical shape in both instruments; their vibration recipes have no significant suppression of the even harmonics. The trumpet sounds more "brassy" than the French horn, and we can see the reason for this difference in the vibration recipes: The amplitudes of the higher harmonics fall off more rapidly for the French horn than for the trumpet. It is the presence of these higher harmonics in the trumpet's vibration recipe that produces its distinctive, brassy sound.

Summary

In the examples of standing waves that we have looked at in this chapter, we have found many similarities to standing waves on strings and in organ pipes. The development of musical instruments (all of which take advantage of standing waves) has occurred over many years using trial and error methods. In recent years, scientists have studied the sound produced by various musical instruments and have come to realize how sensitive the designs are in the production of high quality music. We are not interested in all the design details, but it is important to recognize the similarity between all these different examples of standing waves.

A French horn. The bore on this instrument is much longer than the bores on the trumpet and trombone. (Photograph provided courtesy of C.G. Conn, Ltd., Elkhart, IN.)

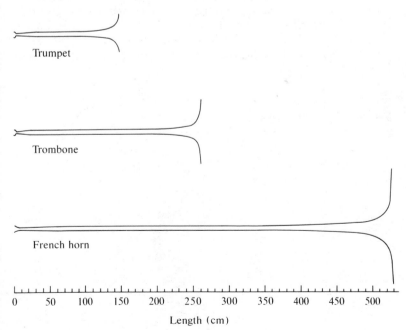

Trumpet

Trombone

French horn

| | | | | | | | | | | | |
|0|50|100|150|200|250|300|350|400|450|500|

Length (cm)

Figure 9.17. The shapes of some brass instruments (The diameter is multiplied by a factor of five compared to the length.) Illustrations based on figures 42, 49–51 (from *Horns, Strings, and Harmony* by Arthur H. Benade. Copyright © 1960 by Educational Services Incorporated. Reprinted by permission of Doubleday & Company, Inc.)

Suggested Readings

The physics of musical instruments is discussed in a number of books. Those that are at about the level of this book include

1. *The Physics of Musical Sound,* by Jess J. Josephs (New York: Van Nostrand and Reinhold Company, 1967).

2. *Physics and the Sound of Music,* by John S. Rigden (New York: John Wiley and Sons, 1977).

3. *Physics and Music, The Science of Musical Sound,* by Harvey E. White, and Donald H. White (Philadelphia: Saunders College/Holt, Reinhart, and Winston, 1980).

4. *Fundamentals of Musical Acoustics,* by Arthur H. Benade (London: Oxford University Press, 1976).

5. *Musical Acoustics,* by Donald E. Hall (Belmont, Calif.: Wadsworth Publishing Company, 1980).

All of these books have quite complete descriptions of sound. That is, they include discussions of vibrations and waves as applied to sound as well as sections on standing waves on strings and in pipes, how the ear works, and how specific instruments work.

A number of *Scientific American* articles on physics and music have been gathered together in a reprint book, *The Physics of Music* (San Francisco: W. H. Freeman and Company, 1978). Articles that apply to topics in this chapter are: "The Physics of Wood Winds," by Arthur H. Benade (October 1960), "The Physics of Brasses," by Arthur H. Benade (July 1973), "The Physics of Violins," by Carleen Maley Hutchins (November 1962), and "The Physics of the Bowed String," by John C. Schelleng (January 1974).

Review Questions

1. Which one of the five diagrams in Figure 9.18 shows the second normal mode of seiching in an east-west direction for the bay in Figure 9.3?

   a. A

   b. B

   c. C

   d. D

   e. E

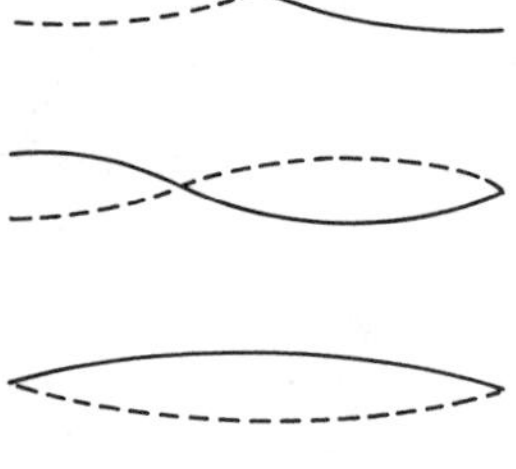

Figure 9.18. For use with review question 1

2. Which of the following similarities between an east-west seich, as shown in Figure 9.3, and a standing sound wave in a closed organ pipe is correct?

    a. They are both built out of traveling longitudinal waves.

    b. They are both built out of traveling transverse waves.

    c. They both have ends that are alike.

    d. They both have ends that are different.

    e. Both systems can have even as well as odd harmonics.

3. A lock in the Panama Canal has a seich in it while the water is rising. What happens to the seich?

    a. It dies out because the lock will have a resonant frequency at only one water depth.

    b. The distance between nodes will increase as the water gets deeper.

    c. The distance between nodes will decrease because the waves will be traveling faster.

    d. The frequency of the seich will increase because the waves travel faster.

    e. The frequency of the seich will decrease because the waves travel faster.

4. Suppose a fundamental mode seich occurred in Lake Erie. What would be the period of such a seich? (Lake Erie is approximately 64 meters deep and 400,000 meters long. We can assume the ends of the lake act as steep sides.)

    a. 10 minutes, 27 sec (627 sec)

    b. 1 hour, 44 minutes (6,250 sec)

    c. 8 hours, 46 minutes (32,000 sec)

    d. 13 hours, 53 minutes (50,000 sec)

    e. 296 days, 7 hours (25.6 million sec)

5. The lowest note on the violin is G_3, with a frequency of 196 Hz. The length of this string is approximately ½ meter. Therefore, the speed of the waves traveling along this string is approximately

a. 98 meters per second.

b. 196 meters per second.

c. 330 meters per second.

d. 392 meters per second.

e. 800 meters per second.

6. The speed of sound is approximately 330 meters per second, and the frequency of the note E_5 is approximately 660 Hz. Thus, the effective length of the clarinet's bore when playing this note is about

a. ⅛ meter.

b. ¼ meter.

c. ½ meter.

d. 1 meter.

e. 2 meters.

7. A ram's horn was used long ago as a ceremonial instrument. Since it essentially has a conical bore that is closed at the small end, we expect it to

a. produce only soft tones.

b. produce only loud tones.

c. have only odd harmonics in its vibration recipe.

d. have only even harmonics in its vibration recipe.

e. have all harmonics in its vibration recipe.

8. We can compare the violin to the clarinet. The functions of the bow, strings, and fingers in playing the violin compare directly to the following sets of objects for playing the clarinet:

a. reed, air column, and finger holes.

b. bore, lips, and bell.

c. bell, bore, and air column.

d. bore, reed, and bell.

e. lips, bell, and finger holes.

10

Light as a Wave

Important Concepts

- Light carries energy and travels in straight lines.

- Light is that small section of the electromagnetic spectrum that is visible.

- All electromagnetic waves travel at the same speed in a vacuum. Matter is a dispersive medium for light; the long wavelength red waves travel faster than the short wavelength blue ones.

- The three additive primary colors are red, green, and blue, and the three subtractive primary colors are cyan, magenta, and yellow.

- Three types of light sources are incandescent materials, gas discharge tubes, and fluorescent materials.

At the beginning of the book, I stated that light was one of the waves we would be studying. Just as with sound waves, we cannot actually see waves of light. Instead, we know that light is a wave because of the way it behaves. The discussion in this chapter will show that it is plausible for light to be considered a wave. In later chapters when we look at additional ways that waves behave and see that both light and sound act in these ways, then we will be more convinced that light and sound are waves.

Description of Light Waves

One of the first properties of waves we described was the ability to transport energy. Thus, if we want to show that light is a wave, we must first demonstrate that it does carry energy. It is easy to demonstrate this ability by using a pinhole camera. Of course, any camera would do, but a pinhole camera also can demonstrate that light travels in straight lines. Figure 10.1 shows how light from an object (the tree) passes through the small opening on the front of the box (the pinhole) and strikes the film in the back of the camera. Note that light from the top of the tree strikes the bottom of the film, and light from the bottom of the tree strikes the top of the film. Thus, an inverted and left-right reversed image of the tree is formed on the film. (Note, however, that the image on the film has the same shape as the object. By rotating the film after exposure, the picture will have the same orientation as the original object.) The image is recorded permanently on the film because the energy carried by each bit of light initiates a chemical reaction in the film where the light strikes it. This chemical reaction occurs in a light-sensitive layer of the film called the *emulsion*. The other layer of the film is a transparent celluloid that is just used to support the emulsion.

Figure 10.1. The formation of an image on the film of a pinhole camera

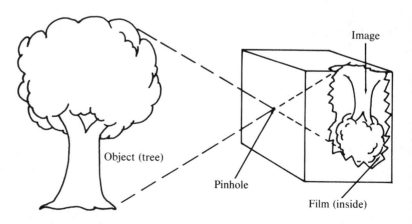

Image

Object (tree)

Pinhole

Film (inside)

The light reaching the film only starts the chemical reaction to record the photographic image. Developing the film completes the process. During development, there are basically three steps in which chemicals are used (for black-and-white film) to finish what the light started. In the first step, we use the *developer*, itself. The developer causes the same chemical reaction to occur as when the light hits the film. The reaction with the chemical is much faster, however. In this chemical reaction, metallic silver is deposited on the film, and the film turns black. The second step in the development process uses a chemical called a *stop bath* for the obvious reason that it stops the chemical reaction started by the light and continued by the developer. Finally, the film is cleaned of its residual light-sensitive emulsion by use of a chemical called a *fixer* so that the image is stabilized (or fixed) on the film.

The entire exposure and development process leaves the film blackened wherever the light struck it. Thus, a light sky appears black on the film, and a dark mountain appears light. Because of this reversal, the film is called a negative. To obtain a positive print (where the image will look just like the original object), the process is repeated. This time, however, we use an emulsion-coated paper. We expose it to a uniform light source but use the negative as a mask. Again, the light striking the emulsion causes the emulsion to blacken. But now, the dark image of the sky on the negative prevents light from reaching the emulsion on the paper, so the sky is light on the developed print. Similar reasoning tells us that the light-appearing mountain on the negative film will allow light to expose the paper, so it will be dark on the print. As with the film, the chemical reaction is started with the energy carried by the light and must be completed with a similar developing process. (For an "instant" camera like those marketed originally by Polaroid, the entire process of producing a positive print after the negative has been exposed is combined into one operation. As the film is pulled or ejected from the camera, chemicals are released automatically to complete development.) Thus, two important properties of light are demonstrated by using a pinhole camera to record a photographic image: Light carries energy, and light travels along straight paths.

Sometimes, light and sound waves act very differently. For example, when you are talking to people who are around a corner, you may hear them quite easily, but you cannot see them. This phenomenon means the sound is bending around the corner, but the light is not. As we will see later, waves with long wavelengths bend around corners more easily than short wavelength waves. And audible sound waves are much longer than visible light waves. In addition, light waves travel much faster than sound or water waves, and this difference in speed also produces apparent differences in behavior. The sizes of the wavelengths; frequencies; and speeds of visible light, audible sound, and water waves are compared in Table 10.

Table 10.   A comparison of light, sound, and water waves

Wave Property	Typical Values of Wave Property for		
	Audible Sound	Water	Visible Light
Speed (meters/sec)	330	3*	300,000,000
(miles/hr)	740	7	671,000,000
Wavelength range (meters)	.02–16	.02–250	.0000004–.0000007
Frequency range (Hz)	20–20,000	.06–5	430 trillion–750 trillion

*Deep water waves with λ = 20 ft.

For light, the speed is so great that its propagation seems instantaneous for most situations. This assumption is not true, however, when astronomical distances are involved. For example, it takes light over 8 minutes to reach the Earth from the sun. Thus, if the sun suddenly were extinguished, people on Earth would not know about it until 8 minutes later. Light from a distant star travels for billions of years between the time it is emitted by the star and the time we observe it.

The speed of light first was determined in 1676 by Olaf Roemer, a Danish astronomer. He used the light reflected from one of the moons of the planet Jupiter. Jupiter has four moons that can be seen easily with a small telescope. Each of them revolves around Jupiter in the same way our moon revolves around the Earth; that is, they move on circular paths. As we view them from the Earth, however, we see the circular orbits of each moon on edge, so each moon seems to be moving along a straight line that passes through Jupiter. Also, during each revolution, the moon is eclipsed (that is, hidden) by the planet. Using one of these moons, Roemer predicted the times during an entire year when the moon would reappear after eclipse. He found, however, that the moon consistently reappeared too early when the Earth was close to Jupiter and was late in reappearing when the Earth was far from Jupiter. He correctly deduced that when the distance between the planets was small, the light made the trip in a shorter period of time, causing the moon to reappear too early.

Just as with the other waves we have studied, the speed of light depends upon the medium through which it moves. Unlike other waves, however, light does not require a medium; it can travel through a perfect vacuum. Light travels as a non-dispersive wave through a vacuum; that is, all wavelengths travel at the same speed. The speed of light in vacuum is also a natural speed limit in the universe—nothing travels faster. When light passes through any type of matter (gas, liquid, or solid), it slows down and is dispersed. Short wavelengths of light travel slower in matter than long wavelengths. We describe the effect a material has on the speed of light waves in terms of the material's *optical density*. The larger the optical density of a material, the more light waves are slowed down as they travel through it. Since all materials are dispersive, the optical density for any material will vary according to the wavelength of the light.

Light is actually an electromagnetic wave. That is, it is a vibration of electric and magnetic fields that can exist in empty space or inside certain matter. The wide spectrum of electromagnetic waves that exists in nature is shown in Figure 10.2. All of the different types of radiation shown here except light are invisible to humans. From the viewpoint of physics and wave description, however, the only difference among these various types of waves is their wavelengths (or frequencies). As noted earlier, these waves all travel with the same speed through a vacuum. But their behavior within matter can be very different. For example, gamma rays (which have much shorter wavelengths than visible light) will pass through sheets of aluminum; whereas, light (and most of the spectrum of electromagnetic waves) will not. We will limit our further discussions of electro-

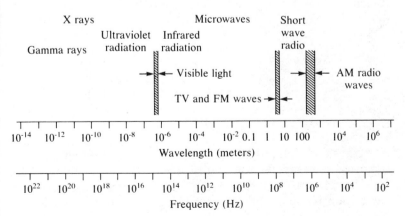

Figure 10.2. Types of waves in the electromagnetic wave spectrum

magnetic waves to light. Note that this is quite a limitation because visible light is only a very small part of the entire spectrum. Because light's frequencies and wavelengths are visible, however, they warrant special treatment.

Color

When white light from the sun or an ordinary light bulb passes through a prism, the light separates into the entire spectrum of the colors of visible light. (See Plate 1 in the color section.) The prism does not create these colors; it merely separates them. They all were contained in the incident white light. The different colors are light waves of different wavelengths. Red light has the longest wavelength (and, therefore, the lowest frequency), and violet light has the shortest wavelength (and highest frequency). (We often will talk about the spectrum extending from red to blue, but when we do so, we are including indigo and violet in the general color of blue.) In between these extremes, we have all the colors of the rainbow. An easy way to remember the order of the colors in the spectrum is to use their first letters to spell a mythical name: *ROY G BIV*. As shown in Table 11, the letters in this name give the various colors of the spectrum from longest wavelength to shortest wavelength.

Table 11. The colors of the spectrum

Color	Frequency (Hz)	Wavelength* (nm)
Red	426 trillion	700
Orange	458 trillion	650
Yellow	497 trillion	600
Green	542 trillion	550
Blue	662 trillion	450
(Indigo	693 trillion	430)
Violet	745 trillion	400

*Wavelength in vacuum or air

Note: Indigo is included in this list of colors in the spectrum to aid in creating the mnemonic *ROY G BIV*.

Waves of visible light are all very short; the approximate range of their wavelengths is from about 400 nm to 700 nm. The "nm" stands for *nanometer* (nan'o • meter), and one nanometer is one-billionth of a meter. Thus, the length of a light wave of any color is less than a millionth of a meter. The range of wave-lengths or frequencies of visible light is also very small: There is less than a factor of two between the smallest and largest. This range should be contrasted to the range of audible sound frequencies, which vary by a factor of almost one thousand: 20 Hz to 20,000 Hz.

The color separation when the light passes through the prism occurs because of dispersion within the glass. To get a feeling for the speed and dispersion of light in a typical type of glass, consider a block of glass a million miles thick! It takes light about 8½ seconds to pass through this block, and when the light emerges, the red light is about 15,000 miles in front of the blue. Remember that red light waves are longer than blue ones and travel faster in matter. However, they both travel at the same speed through a vacuum.

We have seen how to separate white light into its constituent colors by passing it through a prism. We can reverse the process by using lights of all the colors of the rainbow. Combining these colors by projecting them together onto a screen gives back the white light. This process of combining lights of two or more colors is termed *color addition*. Note that the final beam of light contains more frequencies of waves than any of the original light sources. In contrast, *color subtraction* occurs when some color (or colors) of light are absorbed from an original beam. In this case, the final light beam contains fewer wave frequencies than the original beam. In each of these processes, the color perceived in the final beam is different than that in the original one. Just as with the perception of sound discussed in Chapter 5, the way we detect and interpret light is very complicated.

We will discuss the physiological and psychological aspects of vision in Chapter 12. Here, we will look at two different sets of primary colors that physicists use to describe the processes of color addition and subtraction. The primary colors are constructed by dividing the visible spectrum into three equal sections. (See Plate 2 in the color section.) When viewed alone, the three sections appear *red*, *green*, and *blue*, respectively, even though each section contains a range of wavelengths. These three sections are the *additive primary colors*. When correct intensities of these colors are added together, we have the entire spectrum, and we see white light. Other colors can be produced from these primaries by using suitable combinations of them. When correct intensities of the blue and green primaries are added together, for example, blue-green light is produced; this color is called *cyan*. By decreasing the intensity of the primary blue light, the resulting color will become greener, and vice versa. Note that cyan contains waves of all frequencies in two-thirds of the visible spectrum. In fact, cyan is just white light that has had the red additive primary subtracted out. For this reason, cyan sometimes is called *minus red*. In the same way, correct intensities of the red and green additive primaries produce *yellow*, and correct intensities of blue and red produce *magenta*. Following the reasoning developed for cyan, we sometimes call yellow *minus blue* and magenta *minus green*. Since these latter three colors are obtained by subtracting one-third of the visible spectrum from white light, they are known as the *subtractive primary colors*.

It is important to recognize that these two sets of primary colors are used in different ways. When combining (or adding) two or more sources of light, we use the additive primaries to understand the resulting color. The subtractive primaries are used when the color of a single source of light is being changed. A couple of examples of each of these processes may help to clarify the situation. Plate 3A (in the color section) shows three beams of light illuminating the same screen. The colors of the three beams are blue, green, and magenta, and we wish to know the color displayed on the screen. Since this process involves color addition, we use the additive primaries. We must recognize that the magenta beam of light contains both blue and red. So, we essentially are adding four primary beams: two blue, a green, and a red. But this process is the same as adding a blue and a white because one of the blue beams combines with the green and red to form white. When blue and white are combined, the result is a pale blue color; it is termed a *less-saturated* blue.

A second example of color addition is shown in Plate 3B (in the color section) where blue and yellow beams of light strike a screen at the same location. Following the same procedure as in the previous example, we first recognize that yellow is made up of green and red, and, of course, the addition of blue, green, and red gives white light. Thus, the addition of blue and yellow gives white. This result is not surprising since the yellow is just white light with blue removed. When two colors (such as yellow and blue) combine to give white light, the colors are termed *complementary colors*. Magenta and green is another obvious pair of complementary colors as is the pair cyan and red.

The first example of color subtraction is shown in Plate 4A (in the color section), where yellow light shines through a cyan filter. The yellow light contains both primary green and primary red, and the cyan (minus red) filter absorbs primary red. Thus, on the far side of the filter, we observe only primary green light.

A second example of color subtraction is shown in Plate 4B (in the color section). Here, white light is sent through two filters: magenta and blue. Again, we reason out the resulting color using the primary colors. The magenta (or minus green) filter removes the primary green from the white light. This separation means, of course, that the red and blue primary colors pass through. Now, the blue filter allows only blue light through—both green and red are absorbed (if they are present). Thus, the only color that passes through both filters is blue.

The most extreme case of color subtraction occurs when we send white light through enough filters to remove all of the light. The resulting "color" is black because there is an absence of light. Examples of this situation include sending light through red, green, and blue filters or sending it through blue and yellow filters. In these cases, each filter removes part of the spectrum; combining their effects removes all light. If we try these examples at home or in the lab, we probably will see a small amount of light getting through. Real filters do not work perfectly. Thus, when all wavelengths of light are supposed to be absorbed, some still may be seen.

Color by subtraction is far more prevalent in the everyday world than additive coloring. Whenever we view an object that is not, itself, emitting light, we see the object because of the light it reflects. Tree leaves appear green in the afternoon sun because the white sunlight is partially absorbed by the leaves and partially reflected. A leaf acts in the same way that a green filter acts, except the green light is reflected rather than transmitted. (The color of a leaf is not a true primary green; the chlorophyll contained in the leaf absorbs strongly in both the blue and red parts of the visible spectrum, so the dominant color in the reflected light is green.) In a similar fashion, a red sweater appears red in white light because other colors are absorbed by the pigments in the dyes that colored the yarn. Note that if a red sweater is illuminated with blue light, then the sweater will appear black because no light is reflected. The blue light is absorbed, and there is no red light available to be reflected.

The printing of colored pictures in magazines and books also uses the subtractive primary colors. The original colored picture is photographed three times, each time through a different filter of a subtractive primary color. In this way, the original colors are *separated* into their various components. That is, if a part of the original picture is a blue sky, it will be present in both the cyan and magenta photographs. These three photographs are then printed in the respective primary color, on top of one another to reproduce the full colors of the original picture. Since each subtractive primary color absorbs some of the incident light, the correct combinations of the subtractive primaries will yield the correct original color. For the example of the blue sky, the magenta color will absorb any green from the incident light, and the cyan will absorb any red. The resulting blue is the color of the sky in the original picture. There is one additional refinement to this process. The three primary colors actually cannot be printed directly on top of each other because one color would hide the other. Consequently, each color is printed as an array of very small dots, and the different colors are printed side by side. Better quality pictures (such as those used to reproduce artistic masterpieces in museum guides) use smaller dots. Colored pictures in newspapers are generally of lower quality, and the larger dots can be seen with the naked eye. A magnifying glass or microscope must be used to see the individual dots in higher quality printed material.

Sources of Light

Light waves are emitted by atoms. But atoms are so small that we cannot observe them directly. Thus, we actually cannot see vibrations that produce light waves. The waves of light are not directly visible, either, so you might wonder why we include light in our discussion of waves. The reason (as mentioned earlier) is that light behaves like a wave, and so, our knowledge of waves will enable us to understand the behavior of light. Since light waves carry energy, and since they are produced within atoms, they must receive this energy from the atoms, themselves. During the emission process, an atom gives some of its energy to the light wave it produces. But where does the atom get its energy? We will find that there are several ways for atoms to get energy; the type of light source depends on the source of energy for the atoms.

All matter emits electromagnetic radiation, but this radiation is not all visible, of course. The wavelength of the emitted radiation depends on the temperature of the matter. Higher temperatures produce shorter wavelengths (or higher frequencies). When the temperature of the matter is high enough, the matter will emit visible light, and we term the matter *incandescent*. Thus, the sun is an incandescent object because its matter is hot enough to emit visible light. In the same way, the glowing filament in a light bulb produces its light because of its temperature, so we call the object an incandescent lamp. The atoms of such hot objects are in rapid motion, and rapidly moving atoms have energy that can be transferred to emitted light waves. In addition to the sun (and other stars) and incandescent lamps, this rapid movement of atoms and energy transfer also happens in toasters, ranges, and oven heating elements. The high temperature in the sun and stars is a result of nuclear reactions occurring in the centers of these bodies. The heating elements and incandescent lamp filaments get hot when an electric current passes through them. (We find that whenever an electric current passes through a solid material, the material's temperature rises.) In addition to visible light, these sources also emit heat as infrared radiation and may emit ultraviolet radiation. Thus, the moving atoms emit waves of many wavelengths and frequencies. For incandescent objects, some of the waves lie within the range that humans can see. Figure 10.3 shows the wavelength spectra for the radiation emitted from the sun, a light bulb, and a toaster element. We will see presently that the differences in these spectra cause the objects to have different colors.

In contrast to these incandescent sources, fluorescent materials remain cool while emitting light. The atoms in these materials take in energy from other electromagnetic waves and then give it to the emitted visible light. Most novelty shops sell "black lights" and black light posters. When turned on, the black light appears violet; it is a source of ultraviolet radiation, but it also emits a small amount of visible violet light. The black light

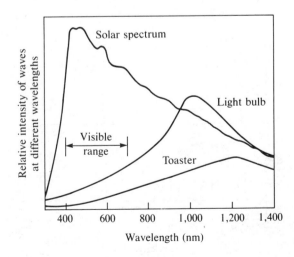

Figure 10.3. Relative amounts of different wavelength waves in the emission spectra of the sun, a light bulb, and a toaster

poster is printed with fluorescent inks that absorb the ultraviolet radiation and use that energy for the light being emitted. Because these posters are actual sources of light, they appear more brilliant than other posters that are illuminated with light originating elsewhere.

A neon sign is an example of another type of light source that remains cool. This type of source is given the generic name *gas discharge tube*. Here, an electric current passes through a gas, and the atoms of the gas absorb energy directly from the electric current. Then, the atoms give up this energy when light is emitted. There is a very distinctive difference between the light emitted by a hot object (such as a filament in an incandescent bulb) and that emitted by a gas discharge tube. In the former case, many wavelengths and frequencies of light waves occur in the emitted light. We say there is a *continuous spectrum* of frequencies or wavelengths of the emitted light. Light from a gas discharge tube, however, has only certain frequencies; different sets of frequencies (and, therefore, wavelengths) occur for different gases. In this case, there is a *discrete spectrum* of emitted light frequencies or wavelengths. Several examples of gas discharge emission spectra are shown in Figure 10.4. Note that in each of these, there is only a limited number of discrete frequencies or wavelengths of light emitted. The small number of distinct frequencies should be contrasted to the continuous range of emitted frequencies for the incandescent objects illustrated in Figure 10.3.

The common fluorescent light bulb combines both the gas discharge tube and fluorescence. There is a coating of material on the inside wall of the glass tube. The tube also contains mer-

Figure 10.4. Emission spectra for three elements. Light is emitted only at the wavelengths where the lines are located.

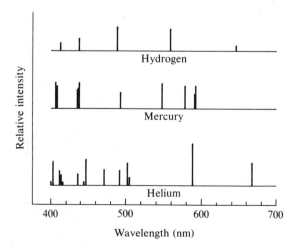

cury vapor, and an electric current is sent through the vapor. The vapor emits a limited number of wavelengths of light. This light, which is not acceptable for indoor lighting, is absorbed by the coating material. The coating fluoresces and gives off a more suitably colored light.

There is a fourth basic type of light source called a laser. It is similar to the gas discharge tube and fluorescence in that it does not operate at high temperatures. However, laser light has several important and unique features. Thus, we postpone a full discussion of lasers until Chapter 15.

Summary

We claim light is a wave, and we have started supporting this claim by noting that it transports energy. The process of photography uses the energy carried by light to initiate a chemical reaction in the emulsion layer of the film. By using a pinhole camera, we also see that light travels in straight lines. The frequency and speed of visible light are much greater than those of water and audible sound waves, and light's wavelengths are much shorter than the wavelengths in water and audible sound waves. Visible light is only a small part of the entire spectrum of electromagnetic waves. All electromagnetic waves travel at the same speed in a vacuum, but they can behave quite differently when encountering matter. Light exhibits dispersion in matter; the long wavelength red light is faster than the short wavelength blue light.

White light is a combination of all colors. When we pass it through a dispersive prism, we get the entire rainbow of constituent colors, each having a different wavelength and frequency. We can remember the order of these colors by using the fanciful mnemonic: ROY G. BIV. Physicists use two sets of primary colors: red, green, and blue for color addition and cyan, magenta, and yellow for color subtraction. The latter process occurs far more often in the world around us because we see most objects by means of the light reflected from them. The light is filtered during the reflection process, eliminating some colors.

Suggested Readings

An excellent book on light, written at the level of this book, is *Light and Color*, by Clarence Rainwater (New York: Golden Press, 1971). This little book is delightful because there are so many pictures, and all of them are in color! Practically everything connected with light and color is covered in an elementary fashion.

The Universe of Light, by Sir William Bragg (New York: Dover Publications, Inc., 1959) is a nice book on many aspects of light. It is written in non-technical language and should be understood easily by readers of this book.

Review Questions

Figure 10.5. For use with review question 1

1. The man in Figure 10.5 is photographed with a pinhole camera. The image on the film looks like which of the following:

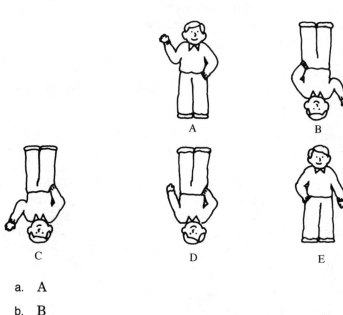

   a. A
   b. B
   c. C
   d. D
   e. E

2. At its closest approach to Earth, the planet Jupiter is 628.7 billion meters from Earth. How long does it take for light to reach the Earth from Jupiter?

    a. 32 sec

    b. 8 minutes, 13 sec (493 sec)

    c. 15 minutes, 37 sec (937 sec)

    d. 22 minutes, 7 sec (1327 sec)

    e. 34 minutes, 56 sec (2095 sec)

3. An organ pipe 2 meters long will contain four complete sound waves of the frequency 660 Hz. About how many green light waves will fit in this pipe?

    a. 3,500

    b. 35,000

    c. 350,000

    d. 3.5 million

    e. 35 million

In each of the following five questions, the colors given are projected onto the same area of a screen. Choose the best answer for a final color from this list: a. white, b. cyan, c. magenta, d. green, and e. red.

4. Red and blue

5. Magenta and green

6. Magenta and yellow

7. Magenta, yellow, and cyan

8. Blue, yellow, and magenta

In each of the following five questions, an initial colored beam of light passes through the indicated filter or filters. Pick the best choice for a final color from this list: a. black (no light), b. red, c. green, d. blue, and e. yellow.

9. A cyan light passes through a green filter.

10. A white light passes through a yellow filter.

11. A magenta light passes through a cyan filter.

12. A green light passes through magenta and cyan filters.

13. A red light passes through yellow and cyan filters.

11

Wave Movements

Important Concepts

- Wavefronts are lines of constant phase along a wave, and rays are paths followed by the energy moving with the wave. Wavefronts and rays are always perpendicular to each other.

- Refraction is the change in direction of a wave due to a change in wave speed.

- When waves cross an interface between media having different speeds, the amount of refraction depends on both the ratio of speeds and the angle of incidence.

- For any case of refraction, the ray is always closer to the normal in the slower medium.

- The incident angle is termed the critical angle when the refracted ray lies along the interface. This situation only can happen when the wave is moving from a slow to fast medium.

- Total internal reflection occurs when a wave moves from a slow medium to a fast one with an incident angle greater than the critical angle.

- The easiest way to determine the behavior of waves reflected or refracted from curved interfaces is to construct successive wavefronts. Using the wavefronts, the rays can be constructed.

- Rays of a wave being focused by a surface or a lens converge toward or diverge from the focal point.

- Two types of images can be formed with surfaces and lenses. A real image is one that can be projected on a screen, while a virtual image only can be observed.

- We can construct a ray diagram with only two rays to find the location and characteristics of an image formed by a simple lens.

- Dispersion occurs when the wave speed depends on the wavelength.

- The frequency shift known as the Doppler effect occurs when the source or detector of waves is moving. Relative motion toward one another causes an increase in the detected frequency.

In this chapter, we will return to the task of building our vocabulary for describing waves. Thus far, we have looked at waves on a string and in a pipe, waves on water surfaces, and sound and light waves. Each of these three sets of waves travels in a different dimension. The waves on a string or in a pipe are one-dimensional (or, we could say they travel in a one-dimensional space) because we need only one number to locate each point along the string or pipe. (This number, for example, could be the distance from one end.) Water waves travel in a two-dimensional space because we need two numbers to locate each point on the surface of the water. (For example, one number tells how far north of an origin the point is, and the other number tells how far east of the origin the point is.) Finally, sound and light waves are three-dimensional because we need three numbers to describe the location of a point. The third number tells how high the point is.

It becomes progressively more difficult for us to describe wave motion in each of these cases. For example, waves traveling on a string can move only to the left or to the right. But waves on the surface of water can be moving in many directions at the same time. To help us describe these more complicated waves and the way they move, we use the idea of wavefronts introduced at the end of Chapter 6. However, wavefronts only tell us the shape of waves at different locations. To increase our capability for describing what waves are doing, we will add to our vocabulary the complementary concept of *rays*.

Reflection and refraction are processes where waves change direction; the concepts of wavefronts and rays will be very helpful in understanding why the waves make particular directional changes in these processes. Reflection and refraction are important processes for helping us understand how sound, light, and water waves behave. For example, architects require a thorough knowledge of how sound waves reflect and refract when designing concert halls and auditoriums that have good acoustics. And as we saw at the end of Chapter 6, water waves will change direction as the water depth changes.

In rare cases, the topography of the ocean bottom can cause waves to focus toward a particular point. The resulting waves at such a location can be quite large and destructive and are sometimes totally unexpected because the prevailing ocean waves are not very large. Subsequent surveys of the ocean bottom have shown why such focusing occurred for those particular waves.

Reflection and refraction find wide practical use. In connection with light waves, scientists and engineers who design optical instruments such as microscopes, telescopes, camera lens systems, or binoculars require an understanding of these processes. There are many other applications of wave reflection and refraction; one such application that you may not be aware of is in the search for oil deposits. In such searches, geologists and petroleum engineers set off explosive charges on the earth's surface. These explosions send waves through the earth (just as an earthquake would), and wherever the waves change speed, they are reflected. Sensitive detectors placed on the earth's surface record the times between the explosion and the arrival of the waves reflected from various strata within the earth. Using these times, the scientists and engineers construct models of the earth's composition at the test location. When these models indicate a high probability for finding oil, then drilling is started.

Although most practical applications of reflection and refraction result from using these processes to focus waves, in this chapter we will discuss one application that does not: the modern development of light pipes. They utilize, instead, the concept of total internal reflection. Since the most common application of wave focusing is with light waves, our discussion will include simple lenses.

We also will discuss the Doppler effect, which occurs when the wave source or detector is moving. This effect has nothing to do with reflection or refraction, but the observations associated with it can only be explained using waves. Since we observe these results with both light and sound, this effect gives additional support to our assertion that they are waves.

Wavefronts and Rays

Before we look at some specific examples of refraction and reflection, I want to first discuss two ways of describing two- and three-dimensional waves. The first way uses *wavefronts*, which are lines or surfaces of constant phase. In Chapter 6, we described wavefronts using water waves. There, we noted that under certain conditions, we can see long, smooth lines of wave crests on the surface of the water. These lines may be curved rather than straight, but they are easy to follow with your eye. Since every point on the crest of a wave has the same phase, these lines are wavefronts. Of course, a line along a trough is also a wavefront, and a line that is halfway up the back side of the wave is also a wavefront. (In each of these cases, the phase is the same along the line being used as the wavefront.) It usually is easiest, however, to think about a line along the crest of the wave when we think of a wavefront.

For a three-dimensional wave (such as a sound wave or a light wave), the wavefront is a surface rather than a line. The idea is the same, however: The phase of the wave vibration has the same value everywhere on the wavefront. Because such waves travel in three dimensions, it is very hard to draw pictures of them, but we will describe some wavefronts for different types of three-dimensional waves in a moment.

Figure 11.1   Spherical waves

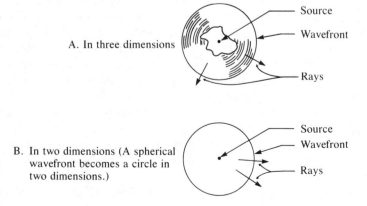

A. In three dimensions

Source

Wavefront

Rays

B. In two dimensions (A spherical wavefront becomes a circle in two dimensions.)

Source

Wavefront

Rays

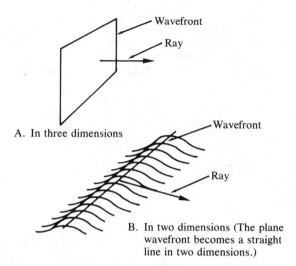

A. In three dimensions

B. In two dimensions (The plane wavefront becomes a straight line in two dimensions.)

Figure 11.2. Plane waves

The other way of describing waves involves *rays*. A ray is simply the path along which a point on the wave moves. We must remember that as the wave passes through a medium, the medium, itself, does not move along with the wave. However, the energy contained in the wave does move, and the paths along which the energy moves are called rays. It also is true that the wave always moves in directions that are perpendicular to its wavefronts. This idea is important because it connects the two ways of describing waves. That is, wavefronts and rays are always perpendicular to one another. Thus, if we can figure out the shape of the wavefronts in a particular situation, then the pattern of the rays follows, or if we know the layout of the rays, then the shapes of the wavefronts follow.

Figures 11.1–11.3 show some different shapes of waves and their movements. Those waves in Figures 11.1 and 11.2 have wide applicability. A *spherical wave* like the ones shown in Figure 11.1 occurs whenever a wave leaves a small source and is still close to it. (It is called a spherical wave because the wavefronts for such a wave—in three dimensions—are spheres. In two dimensions, the wavefronts are circles.

A *plane wave* like the ones shown in Figure 11.2 (for which the wavefronts are planes in three dimensions and straight lines in two dimensions) often is encountered in the study of light. But in any type of wave, a plane wave occurs when all parts of the wave are moving in the same direction. This phenomenon happens when a wave is very far away from its source, that is, thou-

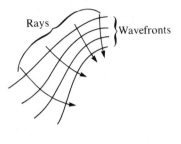

Figure 11.3. Wavefronts and rays for a typical shallow water wave

sands of wavelengths away. As we noted, the spherical and plane waves are given those names because of the shapes of their wavefronts in three dimensions.

Figure 11.3 shows a two-dimensional wave with curved wavefronts and rays. This type of situation might be observed for a shallow water wave moving through an area where the water depth is uneven, causing the wave to change its speed and, therefore, its direction (as we will see in just a few minutes). Note that the wavefronts and rays are always perpendicular to each other wherever they meet.

Refraction

Now, let's consider refraction by looking at a specific case: shallow water waves that move at an angle across an abrupt change in depth. We call the dividing line between the two depths the *interface* between the two regions where the water wave has different speeds. Figure 11.4 shows a series of plane wavefronts on both sides of the interface at one instant of time. The distance between adjacent wavefronts is the distance the wave travels during one period of vibration, that is, one wavelength. Since the wavefronts are closer together on the right-hand side of the interface, it is clear that the wave speed is smaller on that side. Note that as the wave crosses the interface, both the wave speed and the wavelength change, but the wave frequency stays the same. The fact that the wavefronts make a different angle with the interface is due to the change in wave speed.

Focusing our attention on the wavefront that is in the center of the picture and that passes through point *P,* we see that it will take the same amount of time for point *A* on the wavefront to reach the interface as it has taken point *B* to move from the interface to its present location. (Both points *A* and *B* are one wavelength away from the interface.) Since the wave is traveling slower on the right-hand side of the interface, point *B* is closer to the interface than point *A*.

As we have noted, rays are always perpendicular to wavefronts, and since the wavefronts on the two sides of the interface point in different directions, the rays do also. Thus, instead of the wavefront diagram of Figure 11.4, we could draw a diagram showing only the rays of the waves involved. This diagram is shown in Figure 11.5 along with another line that is perpendicular or *normal* to the interface and is used to show the interface's orientation. This normal is used to describe the interface when we are talking in terms of rays because the angle between the

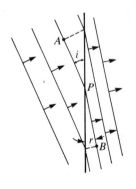

Figure 11.4. The wavefronts on each side of an interface. The wave moves slower on the right side of the interface.

ray (of a wave) and the normal is the same as the angle between the wavefront and the interface.

In Figures 11.4 and 11.5, these two equal angles for the incident wave both are named i. For the wave approaching the interface, this angle is called the *angle of incidence* and, appropriately, the angle for the refracted wave (marked r in each figure) is the *angle of refraction*. These diagrams, of course, show the situation for only one specific case, (that is, one specific angle of incidence and the corresponding angle of refraction).

The amount of refraction (that is, the magnitude of the change in wave direction) depends on two things. First, the change in wave speed is important. As the difference in wave speed increases, more refraction will occur. Conversely, of course, if the change in speed disappears, so too will refraction. The angle of incidence is also a factor. Waves striking the interface head-on will not change direction. As the angle of incidence increases, the amount of directional change also increases. This phenomenon is indicated in Figure 11.6, which shows ray diagrams for three different angles of incidence. The wave speeds on each side of the interface are the same in each case.

However, for a given situation (that is, wave speeds on each side of the interface and incident angle are fixed), the refraction always will be the same. Also, when we send a wave in exactly the opposite direction, the wave will follow the same path. This idea is illustrated in Figure 11.7, where the ray of a wave moving from left to right across an interface is shown in A. For this case, the incident wave moves through the left medium, and the refracted wave moves through the right medium. In picture B, the wave is moving from right to left across the same interface. Thus, the incident wave, in this case, is moving through the

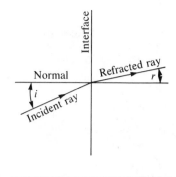

Figure 11.5. The rays of the waves shown in Figure 11.4.

Figure 11.6. Ray diagrams for different angles of incidence.

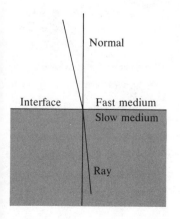
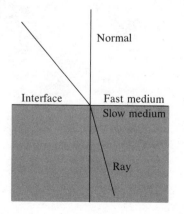
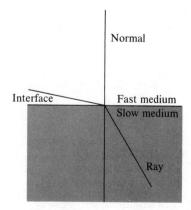

Figure 11.7. Waves moving across the same interface in opposite directions follow the same path.

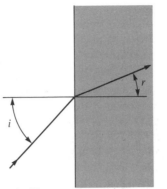
A. Wave moving to right

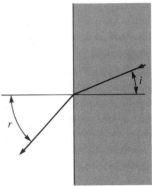
B. Wave moving to left

right medium, and the refracted wave is moving through the left medium. Also, the angle of incidence for the wave in case B has been set equal to the angle of refraction for case A. Since the wave traveling in the opposite direction starts out along the same path, the angle of refraction for case B is the same as the angle of incidence for case A.

In the refraction described here, there is an abrupt change in the wave's speed as the wave crosses the interface. This type of

Refraction of water waves that change speed as they move into a region of different water depth (PSSC Physics, 2nd edition, 1965; D.C. Heath and Company with Education Development Center, Inc. Newton, MA)

A

B

C

situation also occurs when light waves move across an interface between air and water or air and glass, for example. A different type of behavior occurs when the wave speed changes gradually, as in the case of shallow water waves moving in toward the shore over a gently rising sea floor. This behavior was described in Chapter 6, where we saw that as the water depth decreases, the wave speed does also. This reaction produces curved rays rather than ones that are bent sharply. This type of situation was the basis for Figure 11.3. When ocean waves show such a pattern, we can deduce the water depth under the waves. Where the wavefronts are close together, the waves are moving slowly, so the water is shallow. The water is deep where the wavefronts are far apart.

The reflection and refraction of a light beam incident on a piece of glass. The incident beam is coming from the upper left. The refracted beam continues downward, and the reflected beam moves toward the upper right. The angle of incidence increases from A to C.

Reflection

In the previous section, I discussed refraction of two-dimensional waves as they passed across a straight interface. Now, I want to look at how waves are reflected from such an interface (or from a straight barrier). As with refraction, we can consider either the wavefront picture or the ray picture. In discussing the wavefront picture, we recognize that because both the reflected wave and the incident wave move in the same medium, they have the same speed. Because they have the same speed, the angle of reflection is always equal to the angle of incidence. We can see that this statement is true by using the same type of reasoning that we used to determine how the refracted wave changed its direction. Thus, we note that it will take the same amount of time for point A on the wavefront in Figure 11.8 to reach the interface as it took for point B to move from the interface to its present position. Using geometry, we can show that this relationship means the wavefronts of the incident and reflected waves make the same angle with the interface. In Figure 11.8, the angle of incidence is called i and r is the angle of reflection.

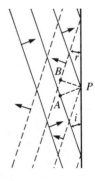

Figure 11.8. The wavefronts in both the incident and reflected waves for a straight barrier

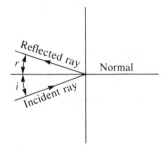

Figure 11.9. The rays of the waves shown in Figure 11.8.

These two angles are always equal. It follows, of course, that the rays describing these waves also makes the same angles with the normal to the interface. The ray diagram corresponding to Figure 11.8 is shown in Figure 11.9; the angles are labeled the same in both figures.

It is helpful to remember here what happened when we considered transverse waves on a string. We found that a reflected wave was generated in a number of different situations: at a fixed end, a free end, a dashpot, and the junction with another string. With two- and three-dimensional waves, we can have situations that are analogous. We will use water waves to illustrate this statement. For their respective waves, a vertical wall in water acts in the same way as a free end of a string; in each case, the wave is reflected without a phase change. A shallow water wave moving across an abrupt change in water depth acts like a transverse wave on a string moving across the joint between two strings. Both a transmitted wave continues on into the second medium, and a reflected wave moves back into the first medium. Also, when the wave speed is faster in the incident medium than in the second medium, there is a 180° phase change upon reflection. Finally, a gently sloping beach acts for water waves in the same way as a perfect dashpot does for waves on a string. No reflected wave is produced in either case. These similarities of behavior for different types of waves are important to recognize. They help us transfer our understanding from simple situations to more complicated ones, and this transfer of knowledge is fundamental to all science.

Total Internal Reflection

We have seen how the direction of a wave changes when refraction occurs. In particular, the ray is bent toward the normal when the wave goes from a fast medium into a slow one. Con-

Reflection of water waves with circular wavefronts from a straight barrier. In A, the waves are approaching the barrier at the bottom of the photograph. The reflected waves are shown in B. (PSSC Physics, 2nd edition, 1965; D.C. Heath and Company with Education Development Center, Inc. Newton, MA)

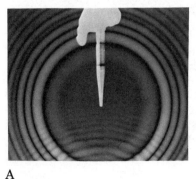

A

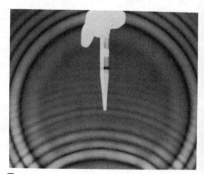

B

versely, of course, the ray is bent away from the normal when the wave moves from a slow medium into a fast medium. We also have noted that the change in direction of the wave will be different for different incident angles.

When waves are moving from a slow medium to a fast one and the angle of incidence becomes larger, the refracted wave will move in a direction that is closer and closer to the interface. At a particular angle called the *critical angle* the refracted wave will move along (that is, parallel to) the interface. (The critical angle depends on the ratio of the wave speeds in the two media.) For an incident wave of any angle between straight ahead and the critical angle, there will be both a refracted wave and a reflected one. The energy carried by the incident wave is shared by the reflected and refracted waves as they leave the interface. An amazing thing happens, however, when the angle of incidence is increased past the critical angle: The refracted wave disappears. This disappearance happens because the angle of refraction is equal to its maximum value when the angle of incidence is equal to the critical angle. Then, when the incident angle increases beyond the critical angle, the refracted angle cannot get any bigger, so the refracted wave disappears. This disappearance means, of course, that all of the energy carried by the incident wave will be carried away by the reflected wave. Since we have only reflection at these larger angles and since this phenomenon happens only when the incident wave is in the slower medium (such as light striking a surface from inside water or glass), we term this phenomenon *total internal reflection*. Figure 11.10 shows ray diagrams for several different cases of a wave moving from a slow medium into a fast one; each case has a different angle of incidence. Note that in the last picture, the incident angle is greater than the critical angle. Thus, total internal reflection is occurring.

It is important to recognize that in order for total internal reflection to occur, the wave must be moving from a slow medium into a fast medium. For water waves, this movement hap-

Total internal reflection of a light beam inside a piece of glass. The incident beam comes in from the left at the top. It passes straight through the left surface and is totally reflected at the upper right and again at the lower right.

Figure 11.10. Waves arriving at an interface at different incident angles, illustrating total internal reflection

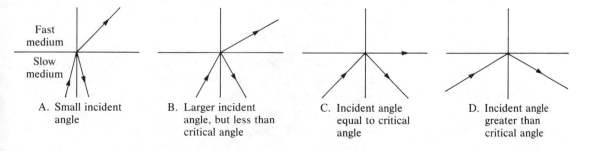

A. Small incident angle

B. Larger incident angle, but less than critical angle

C. Incident angle equal to critical angle

D. Incident angle greater than critical angle

pens when they move from shallow to deep water. To be specific, suppose we have 2-meter wavelength waves in water that is 10 cm deep. (These are very shallow water waves because the depth-to-wavelength ratio is 1/20.) These incident waves would be moving at a speed of about 1 meter per second. Further, let's have these waves move across an underwater drop-off where the water depth suddenly becomes 40 cm. The wave speed doubles to 2 meters per second (and the wavelength also doubles). For these wave speeds, the critical angle is 30°. Thus, for any incident angle larger than 30°, no refracted wave can occur; all of the incident wave's energy will go into the reflected wave.

As another example of total internal reflection, we can look at a situation where light speeds up as it crosses an interface. It is very easy to illustrate this behavior by shining a beam of light from under the water up through the surface. A vertical beam is not refracted at all because it lies along the normal. As we slant the beam away from a vertical position, refraction and reflection both are evident, and the refracted beam is closer to the water's surface than the incident beam. As we increase the angle of incidence through the critical angle (which is about 50° for a water-air interface), the brightness of the reflected beam increases dramatically. This increase in brightness happens, of course, because the energy carried by the incident beam is split when both refraction and reflection occur, but at angles greater than the critical angle, all the incident energy goes into the reflected beam.

The case of light moving from water into air is only one particular example. Two other interesting cases of total internal reflection for light are the brilliant sparkle of a diamond and light pipes. Diamond has a very high optical density, which means that light travels slower in diamond than in most other materials. Consequently, the critical angle for light moving from diamond into air is smaller than for most other materials. By proper cutting, we can make sure that light entering the diamond can only escape through the top surface. In other precious stones or cheap imitations, some of the light entering the stone can leak out the back side. The absence of any leakage of light (by a refracted ray) out of the back of the diamond means it sparkles more brilliantly.

A light pipe is a solid piece of clear plastic rod or fiber. Because the light travels slower inside the material making up the light pipe than in the surrounding air, total internal reflection can—and does—occur at the surface of the light pipe. Thus, light entering one end of a light pipe is trapped inside until it reaches the other end. This phenomenon is illustrated in Figure

Figure 11.11. Ray of light trapped inside a light pipe

11.11, where the light stays inside the fiber even though the fiber is curved. As can be seen, the light remains trapped within the pipe unless there is a sharp bend or kink in the material. If there should be a kink in the pipe, the angle of incidence for the light might fall below the critical angle, allowing some refraction to take place. This situation is shown in Figure 11.12, but we must realize that this refraction is an abnormality for a light pipe.

Figure 11.12. Ray of light in a light pipe. The ray escapes at a kink.

Light pipes have two important applications. Light pipes have been used for a number of years to carry images over short distances. This application is commonly used by doctors, for example, to look at certain parts of the body from the inside. A very small light pipe can be inserted into a vein in the arm, pushed through the vein into the heart, and used by a doctor to inspect the inside of the heart. Or one end of a light pipe can be swallowed by a patient, and the doctor can observe the inside of the stomach. In both these cases and, indeed, anytime that an image is being transmitted along a light pipe, the pipe, itself, will consist of many tiny, individual fibers. The arrangement of the fibers is exactly the same at each end of the light pipe. Each fiber carries only the color and brightness of the light that illuminates the end of the fiber. Since the fiber arrangement is the same at each end, the pattern of light, dark, and color that appears at one end will be reproduced at the other end.

An entirely new use for light pipes has emerged recently with the advent of laser communications. In this application, a

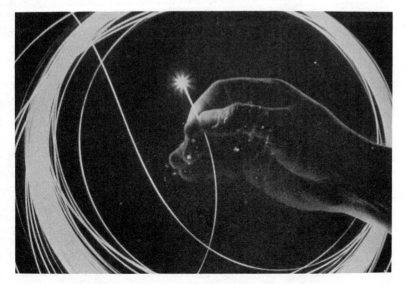

A coil of solid plastic fiber that serves as a light pipe. The light shining from the end of the fiber entered the fiber at its other end. (Courtesy of Ohio Bell Telephone Company)

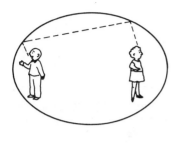

Figure 11.13. A whispering gallery

laser is used to send a light signal into one end of a light pipe. Just as in the previous application, the light signals are carried along tiny fibers. The light signal acts very much like an electrical signal inside a wire. The extremely high frequency of light allows the light to carry a great deal of information. Thus, light pipes are replacing wires within major metropolitan areas where the number of telephone calls is overloading the existing system. Each light fiber can carry more phone conversations that the wire it replaces.

Focusing of Waves

By using curved surfaces for interfaces, it is possible to focus waves. Waves are focused either when they are reflected back from curved interfaces or when they are refracted as they cross curved interfaces. When we use this concept of focusing with light waves, we can understand how the images that occur in the use of cameras, projectors, telescopes, microscopes, and other optical instruments are produced. Focusing can occur in all types of waves, of course. Sometimes, when water waves are focused, large waves that erode shorelines or damage structures along the shore are produced. When sound waves are focused, loud sounds may be produced. The echo is a well-known example of sound waves being focused by reflection. Another example occurs in whispering galleries, which are specially designed rooms. The walls in such a room form the surface of an ellipsoid. In use, one person stands at one focus of the ellipsoid and whispers toward the wall, as shown in Figure 11.13. Another person located at the other focus can hear the whispered words, even though no one else in the room can. This phenomenon exists because all of the sound waves leaving the speaker's mouth are reflected off the walls to the other person's ear. (One such path is shown as a dotted line in Figure 11.13.) Thus, the sound waves are focused to the ear of the listener.

Figure 11.14. Interfaces (surfaces)

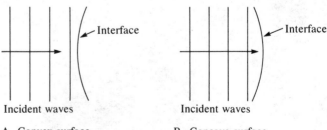

A. Convex surface   B. Concave surface

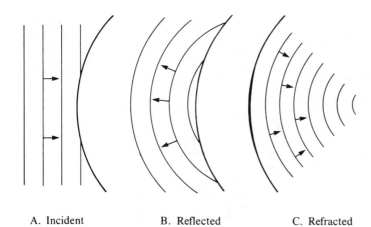

A. Incident   B. Reflected   C. Refracted

Figure 11.15. The wavefronts of the incident, reflected, and refracted waves for the case of a plane wave incident on a convex interface between slow and fast media.

To understand how focusing works, we will consider the behavior of two-dimensional reflected and refracted waves that are formed at a circular interface. Several possibilities exist: The interface either can curve toward or away from the incident wave, as shown in Figures 11.14A and 11.14B. These two types of interfaces are called *convex* and *concave*, respectively. In addition to the two different ways the interface can curve, the wave either can speed up or slow down as it crosses the interface. Thus, there are four different situations that we must consider. Fortunately, just understanding the movement of waves will allow us to determine how the wavefront picture looks for all of these situations.

I will discuss one case in great detail, and then, you should use the same reasoning to see what happens in the other three. I will discuss the case where an incident plane wave is traveling through the faster of the two media on either side of a convex interface. Thus, the refracted wave will travel slower than the incident and reflected waves. The incident wavefronts are shown in Figure 11.15A, and we will look first at the reflected wavefronts in Figure 11.15B. Consider what happens to a plane wave that is reflected from this interface. The wave strikes the center of the interface first, so this part of the reflected wave will be in front of the wave's edges. Thus, the reflected wave is no longer plane; it is now spherical (or circular) with the center of the wave in front. By drawing in the rays for the reflected wave (See Figure 11.16), it is easy to note that the wave is *diverging* from the interface. (We say the wave is diverging because the rays are spreading apart or diverging from one another.) Another way to describe what is happening is to say that the reflected rays

Figure 11.16. The rays of the waves whose wavefronts are shown in Figure 11.15

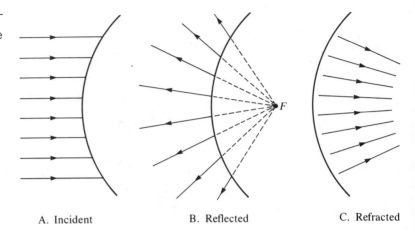

A. Incident   B. Reflected   C. Refracted

appear to be coming from the common point behind the interface, denoted by *F*. (Although dashed lines are drawn in Figure 11.16B connecting the rays with point *F*, no reflected waves occur in this region. They only *seem* to come from the point *F*.) Since all the rays appear to come from this common point, we call it the *focal point*. For the case of light waves, we would

The reflection of light beams off a convex mirror. All the reflected beams appear to come from a single point behind the mirror.

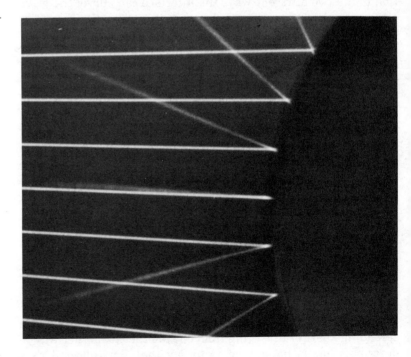

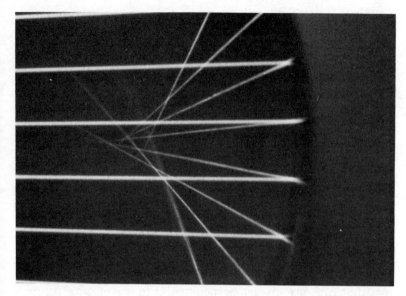

The reflection of light beams off a concave mirror. All the reflected beams almost pass through a common point. There is not a common point because of spherical aberration, which is described in general physics texts as mentioned in the suggested readings at the end of this chapter.

describe this interface as a convex mirror, and we find that reflected waves diverge from a convex mirror.

For the case we are looking at, the refracted wave is traveling slower than the incident wave. Thus, that part of the wavefront that moves across the interface first will lag behind the other parts of the wavefront. For the convex surface being considered here, the center of the wave meets the interface first and is retarded. Since the edges of the wavefronts are ahead of the center, we get the resulting wavefront curvature shown in Figure 11.15C. This curvature means the rays (Figure 11.16C) will pass through a single point, which is also called a focal point. This new focal point does not occur at the same location as the point F in Figure 11.16B. For this situation—and indeed for all but the most unusual ones—the cases of focusing for reflected and refracted waves have different focal points. The location of the focal point for the reflected wave is determined only by the curvature of the surface, while the refracted wave focal point's location also depends on the change in wave speed.

For this refracted wave, we have waves *converging* toward the focal point. The opposite behavior occurs either when the interface is changed to a concave shape or when the wave speeds up as it crosses the interface, rather than slowing down. Just as in the case considered in detail here, these conclusions about the other situations can be obtained most easily by first constructing

Figure 11.17. The wavefronts of the incident, reflected, and refracted waves for the case of a plane wave incident on a concave interface between slow and fast media

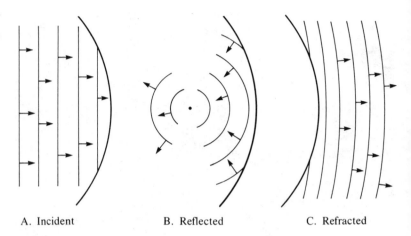

A. Incident   B. Reflected   C. Refracted

the shape of the wavefronts using the known information about the wave speed and then determining the ray picture using the fact that rays are always perpendicular to wavefronts.

Without discussing them in detail, we show the wavefront and ray diagrams for a concave interface in Figures 11.17 and 11.18. Note that the only change from Figures 11.15 and 11.16 is in the shape of the interface; the refracted wave still moves slower than the incident or reflected ones. For the case shown in Figures 11.17 and 11.18, the reflected wave converges and the refracted wave diverges. As mentioned previously, whether a reflected or refracted wave diverges or converges can be figured out easily if one understands the reasoning. On the other hand, it can get very confusing if we just try to memorize all the possible cases.

Simple Lenses

The most common application of refraction at curved interfaces occurs when light passes through glass lenses. In most optical instruments, there are complex lens systems for magnifying or focusing images. These systems are composed of many individual elements, and often, the elements are simple lenses. Also, the behavior of a complex system is usually very similar to that of a simple lens; the complex system just performs better than the simple lens does. Thus, by studying simple lenses, we can learn something about the operation of different types of optical instruments.

We will look at what happens to light waves as they pass through simple lenses; we will use an approach similar to that

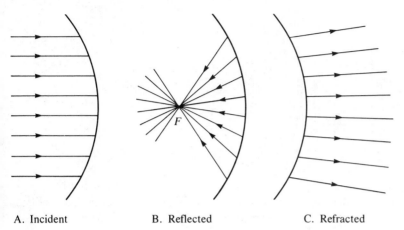

A. Incident   B. Reflected   C. Refracted

Figure 11.18. The rays of the waves whose wavefronts are shown in Figure 11.17

used in the last section. There, we found that it is easiest to look first at the behavior of wavefronts and then to deduce the behavior of the associated rays. It is easier for us to visualize the rays rather than the wavefronts for light, probably because we can construct thin beams of light that act like rays. Even though it is easier to *visualize* the rays, it is often easier to *understand* light behavior using wavefronts, so we must keep the connection between the two firmly in mind.

A good example appears in Figure 11.19, where three thin light beams (acting like rays) pass through a glass lens. Each surface of this lens is convex (that is, bulges out), so the lens shape is termed *biconvex*. It is easy to see that the lens in Figure 11.19 is a converging lens because all three beams pass through a common point. As before, we call this point the focal point.

To gain an understanding of why the beams converge, we look at what happens to wavefronts of a wide light beam as it passes through the lens. (See Figure 11.20.) Notice that since light travels slower in glass than in air, the wavefronts are closer together inside the glass. More importantly, the light

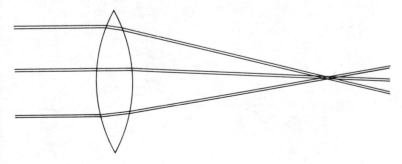

Figure 11.19. Three light beams refracted by a biconvex glass lens

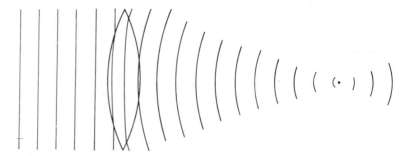

Figure 11.20. Wavefronts for a light beam passing through a biconvex glass lens

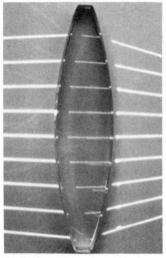

Light beams passing through a converging lens

passing through the thicker central region lags behind that passing through the thin edges of the lens. The curvature of the front edge of the lens causes wavefronts inside the lens to be bent, and the bending is reinforced by the curvature of the second surface. When the rays are drawn perpendicular to the wavefronts of Figure 11.20, they follow paths similar to the ones the thin beams follow in Figure 11.19. Note that a change in wave speed is needed for refraction to occur. Thus, the light is not refracted inside the lens; it changes direction only at each surface.

The important features of the lens in Figures 11.19 and 11.20 are that it is thicker at its center than its edges and that the light travels more slowly in the glass than in the surrounding air. Three other possibilities exist. First, a lens of the same shape could be made from a material in which light travels faster than in the surrounding medium. Such a situation occurs, for example, for an air lens in water. We can construct this type of lens by having a thin glass or plastic container of the appropriate shape that is filled with air and submerged under the water. The behaviors of the wavefronts and rays for light passing through such a lens are shown in Figures 11.21A and 11.21B, respectively. Note that the biconvex glass lens in air is a

Figure 11.21. Light passing through a biconvex air lens in water

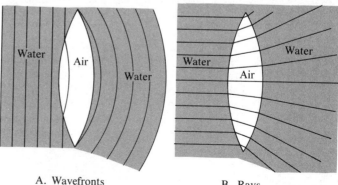

A. Wavefronts   B. Rays

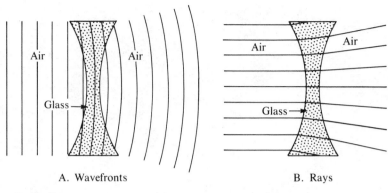

A. Wavefronts   B. Rays

Figure 11.22. Light passing through a biconcave glass lens in air

converging lens, while the same shaped air lens in water is a diverging lens. The difference in behavior of the light is due to the fact that the wave travels slower inside the glass than in the surrounding air and faster inside the air than in the surrounding water.

The other two possibilities occur when we use a lens that is thinner in the middle than at its edges. If both sides of the lens are indented, we term it a *biconcave lens*. But this is only one way of making a lens thicker at its edges; I stress again: The important factor is that the lens is thinner at its center than at its edges. As with the biconvex lens, we can have the light moving either faster or slower inside the lens than in the surrounding medium. These two possibilities are shown in Figures 11.22 and 11.23. In each case, both the wavefront and ray diagrams are shown.

Images

The major reason we use lenses with light is to form images. We saw earlier how light passing through a very small opening (a pinhole) will form an image on film. We saw that the image forms because light arriving at one spot on the film originates from only one spot on the object. A similar situation occurs when we use a lens. That is, light that leaves one spot on an object and passes through a lens can be focused onto a single point on a screen (or a film). If this situation happens for all points of the object, then we have an image produced on the screen.

A diagram of this situation is shown in Figure 11.24. The object and its image on a screen are shown along with a few of the light rays leaving the corner of the flag, passing through the lens, and being focused to the corresponding point in the image on the screen. Whenever an image (such as this one) is viewed using a screen, we call it a *real image*. In contrast, there are

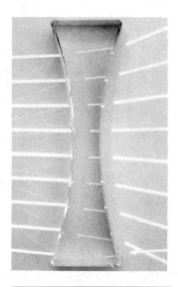

Light beams passing through a diverging lens (The one beam at the bottom is refracted by the first surface of the lens and then strikes the bottom surface of the lens and is reflected upward.)

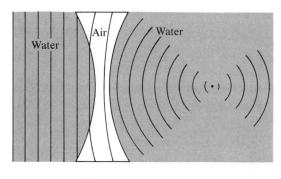

A. Wavefronts

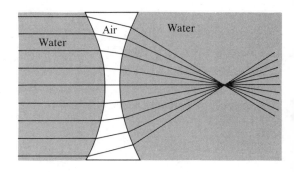

B. Rays

Figure 11.23. Light passing through a biconcave air lens in water

other situations for which the image can only be seen if we look into the lens. This situation always happens when we use a diverging lens to observe an object. Suppose, for example, we hold a diverging lens in front of a penny and look into the lens. We will see the image of the penny; the image will be smaller and will appear behind the lens. there is no way to have this image appear on a screen, so we call it a *virtual image*.

We have seen that a diverging lens can produce a virtual image and that a converging lens can produce a real image. Diverging lenses always produce virtual images of objects. On the other hand, a converging lens can produce both real images and virtual images. In addition to the case we just looked at, we will get a virtual image when a converging lens is held close to the object. If we wish to inspect a small object, for example, we may use a magnifying glass to make the object look bigger. This is an example of a converging lens producing a virtual image. Thus, we find that both types of images can be produced by con-

Figure 11.24. An object and its image

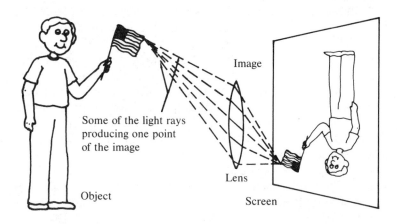

Image

Some of the light rays producing one point of the image

Lens

Object

Screen

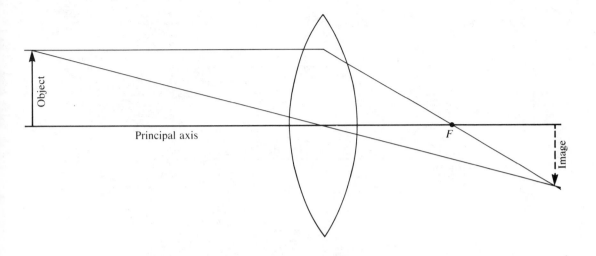

Figure 11.25. A ray diagram for a converging lens

verging lenses, but only virtual images will occur for diverging lenses.

It is relatively easy to construct a ray diagram to determine the location and characteristics of an image. It is easier to perform this construction if we use a simpler object than that shown in Figure 11.24; an arrow is one such simple object. The procedure for this construction is shown in Figures 11.25 and 11.26 for a converging and diverging lens, respectively. In each case, only two rays are needed for the construction. The converging lens case is a little easier, so I shall discuss it first. The object arrow has its tail on the *principal axis*, which is an imaginary line running through the center of the lens. The point on this axis marked F is the focal point of the lens. Its location (that is, the distance from the lens) depends on the shape of the lens and the differences in light speeds inside and outside the lens. Thus, if we have a given lens, then we know (or can determine) the location of the focal point, F. (One way to determine the distance of the focal point from the lens is to shine light from a distant source—such as the sun—through the lens onto a screen. We then change the distance between the lens and the screen until we find the smallest point of light on the screen. That is the location of the focal point for that lens.)

Since we know the location of the focal point, we easily can draw the two rays shown in Figure 11.25. Each ray leaves the tip of the arrow and goes through the lens. The image of the arrow tip is located at the point where these two rays intersect. The first ray goes straight through the center of the lens. The other ray leaves the object in a direction parallel to the principal axis.

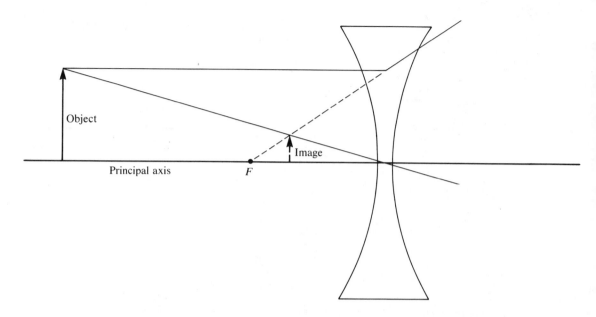

Figure 11.26. A ray diagram for a diverging lens

This ray continues to a point halfway through the lens, changes direction, and goes through the focal point to intersect with the first ray. In this way, we have constructed the image of the arrow tip. The images of all other points on the arrow could be constructed in the same way, but that process is unnecessary. It is sufficient to draw in the rest of the arrow between this position of the tip and the principal axis. Note that the image is real (we could put a screen at its location and allow the rays to fall on it) and inverted (that is, upside-down compared to the object).

Both of the rays we used are only approximately correct. For ease of construction, we drew them as simple straight-line segments. They actually change direction at each surface of the lens; refraction occurs only where the wave speed changes. In spite of these simplifications, this method gives a good picture of image formation by a converging lens.

The construction for a diverging lens is similar but has one complication. In Figure 11.26, we see the same two rays leaving the tip of the object arrow. As before, the first one passes straight through the center of the lens. The second ray is different, however. In this case, the focal point is on the same side of the lens as the object. Thus, the second ray does not bend and go through the focal point. Instead, the ray diverges and appears to come from *F*. For this reason, the dashed line is drawn from *F* to a midpoint of the lens. Note that the intersection of these two rays also occurs on the same side of the lens as the object; this intersection is the location of the virtual image for this lens. Also, note that the image is upright (that is, not inverted).

In each case considered, we could draw many more rays leaving the tip of the object arrow. All of the rays passing through the converging lens would bend and go through the arrow tip's image point. They would appear similar to those shown in Figure 11.24 for the corner of the flag. For the diverging lens, all of the rays leaving the arrow tip and passing through the lens would bend and point directly away from the image of the arrow tip.

A ray diagram for a converging lens with the object closer to the lens (on one side) than the focal point (on the other side) would yield a virtual image because the rays intersect on the same side of the lens as the object. In order to find the intersecting point, the rays have to be extended backward. You should try to construct this diagram! You will find that the image is bigger than the object and is upright.

As we will see in the next chapter, lenses are used to correct visual defects. The type of lens used in a person's glasses depends upon the type of defect being corrected. Anyone wearing glasses, however, constantly views a virtual image of the world around him.

It is also possible to produce images using mirrors. Whenever you look at your reflection in a mirror, you are observing a virtual image of yourself. Since the image is behind the mirror, it is impossible to put a screen at the image's location, so it is a virtual image. Real images can be formed by concave surfaces. An astronomical telescope often will incorporate a concave mirror as the first element in its lens system. Convex mirrors are used as supplementary rear-view mirrors on trucks and automobiles because these mirrors provide a wide-angle view of the region behind the vehicle. We saw in Figures 11.15–11.18 that a convex surface is a diverging mirror and a concave surface is a converging mirror.

Dispersion

We have stressed the idea that the speed of a wave is determined by the medium through which the wave travels. Thus, the wave speed usually will change when the medium is changed. Examples of this phenomenon that we have discussed include

1. A transverse wave on a string slows down when the string is loosened or made heavier.

2. Shallow water waves slow down when the water depth decreases.

3. Sound waves in air slow down when the temperature is lowered.

None of these cases mentioned different speeds for different sizes of wavelengths. In contrast, we noted that for deep water waves and for light waves in matter, the speed depends upon the wavelength (or frequency) of the waves. In particular, we saw that for both deep water waves and light waves in matter, the long wavelength waves travel faster than those with short wavelengths. This effect—where the wave speed depends upon the wavelength—is termed *dispersion*. The speed of the waves still depends upon the medium, of course; so, when this effect happens, we have a *dispersive medium* (that is, one in which dispersion takes place). In the first three cases cited above, the waves are moving through non-dispersive media.

As noted, long wavelength waves travel faster than short wavelength ones for both of the dispersion cases mentioned. For other cases of dispersion, however, we can have the opposite effect. That is, the short wavelength waves travel faster. We noted in Chapter 4 that this effect occurs for transverse waves on stiff strings like piano strings. Water waves also show this type of dispersion; it occurs in the surface-tension waves, which were mentioned in Chapter 6 but which we decided not to discuss in detail. Even though different waves move at different speeds in dispersive media, the speed of each wave is still equal to the product of its frequency and wavelength: $v = f\lambda$.

Several effects occur for waves traveling in dispersive media. The first of these shows up when waves are refracted as they cross an interface. If at least one of the two media on each side of the interface is dispersive, then the change in wave speed as the waves cross the boundary will depend upon the wavelength. This statement means, in turn, that the angle of refraction also will depend upon the wavelength. Thus, if waves of different wavelengths approach the interface with the same angle of incidence, then they will leave the interface with different angles of refraction. We have seen an excellent example of this phenomenon in our discussion of color in light. The white light beam was broken into its constituent colors as it passed through a prism because of this effect. Refraction with dispersion occurred at both surfaces of the prism, and the results were cumulative. Since the shorter blue light waves have a greater change in speed when they cross the interface from air to glass, they will show a greater refraction than the longer red light waves.

We introduced a marching soldier metaphor in Chapter 6, and the metaphor is useful here also. We picture a beam of white light as a file of marching soldiers with various heights. The tall soldiers with long legs represent the red light waves in the beam while the blue light waves are associated with short soldiers having short legs. As long as the file marches on a smooth, level road, all soldiers travel at the same speed. The short-legged soldiers take more steps to keep up—blue light has a higher frequency! However, when the soldiers march off the road into tall grass, there is a greater retarding effect on the soldiers with short legs. This metaphor serves as a nice example to help us remember that blue, short wavelength light waves slow more than others in matter.

It is difficult to observe different refractive angles with water waves because the surface of the water is so choppy when waves of different wavelengths travel together. Even though it's difficult to see these refractive angles, we could do an experiment in a way similar to the one used to observe refraction of water waves. That is, we could send waves at an angle across an interface where the depth of the water changes abruptly. We would have to make two changes from that previous experiment. First, our wave source would have to produce waves having two or more different wavelengths. The second change is required because we need a dispersive medium on at least one side of the interface; thus, we must have deep water waves on one side. On the other side of the interface, however, we must have shallow water waves; otherwise, the waves would not change their speed as they cross the boundary. In this experiment, we would expect to see the long wavelength waves change their direction more as they cross the interface than the short wavelength ones do. It would not matter whether the waves are moving from deep water to shallow water or from shallow water to deep water; the effect would be the same. To summarize: The effect we would see is that a single incident beam composed of different wavelengths would be separated into different beams, each composed of a single wavelength and each going off in a different direction. This separating (or dispersing) of the different wavelengths of the incident beam, which occurs in the refracted beam, shows one source of the name for this effect.

We readily can observe the second major effect of dispersion with water waves. When waves of different wavelengths travel in the same direction at the same time, the displacement of the water's surface will be determined using the principle of super-position. If the waves of all wavelengths travel at the same speed, then the shape of the surface will move along at the speed of the waves. Dispersion, however, will cause the shape of the surface to change. If the wavelengths that are superposing are not very different, then we will get an effect that is similar to surf beat. That is, in some regions of the ocean, the waves will add together to produce larger waves, while in other regions, the waves will tend to cancel each other out.

Where the waves are bigger, we say we have a group of waves. This group of waves will move along together for awhile before the waves disperse. The speed of the group will not be the same as the speed of each individual wave, however. When long wavelength waves move faster than short wavelength ones, the group will move more slowly than the individual waves. In this case, we can watch a group of waves on the surface, and we can see the individual waves moving forward through the group. That is, we can see an individual wave form at the back edge of the group, and we can see this wave move forward through the group as both the wave and the group move together. The effect is reversed when short wavelength waves move faster than long wavelength ones. That is, we can see individual waves form at the front of the group; then, as the waves and the group both move together, the group moves faster than the individual waves, so it swallows them up and then spits them out behind.

In Chapter 8, we observed that the reason the normal modes of a stiff piano string are not harmonics can be traced to the fact that the waves of different wavelengths do not travel at the same speed. Thus, we have another example of dispersion. While it might not rank with reflection and refraction, dispersion is an important wave property. Dispersion is absolutely necessary for explaining some of the behavior we observe in light and water waves.

The Doppler Effect

When we observe waves of any type, one of their primary prop-erties that we are interested in measuring is frequency. Both the pitch and tone quality of the sounds we hear are determined by the detected frequencies. And although we usually talk in terms of wavelengths when discussing light, we must recognize that color is determined by frequency. Also, sailors have named the waves they see on the basis of the waves' periods (or, of course, the waves' frequencies). Up to now, we have inferred that the detected frequency is identical to that of the source. This equality holds only if the source and detector are stationary. When we observe a wave and either we or the wave source (or both, of course) is moving, we will detect a frequency that is shifted from that emitted by the source. This effect is named for the Austrian physicist Christian Doppler, who first demonstrated it in 1842.

In order to understand why the frequency is shifted for moving sources and detectors, look first at Figure 11.27, which shows a number of circular wavefronts moving away from a stationary source S. Each of these wavefronts was emitted by the source at a different time. Those emitted earlier have bigger diameters because they have traveled for a longer time. Since the source is stationary, the center of each circular wavefront is at the same point. The distance between each adjacent pair of wavefronts is the wavelength, and it is determined by the speed of the wave and the frequency of the source ($\lambda = v/f_s$). As these wavefronts move past the detector D, the detected frequency is the number of wave cycles passing each second. This quantity is the speed of the wave divided by the wavelength.

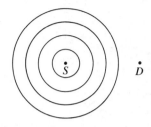

Figure 11.27. Wavefronts of waves emitted by a stationary source, S. D detects the same frequency as that emitted by the source.

$$f_D = \frac{v}{\lambda} = \frac{v}{v/f_s} = f_s \qquad (11.1)$$

Here, the symbols have the meanings: f_D = frequency of wave as detected by the observer, v = speed of wave, λ = wavelength of wave, and f_s = frequency of the wave as emitted by the source. From this exercise, we see that the detected frequency is equal to the source frequency.

The positions of the wavefronts emitted by a source moving to the right are shown in Figure 11.28. Since the source is moving to the right, the wavefronts that are emitted later are the smaller circles and have their centers farther to the right. As a result, the wavefronts are squeezed together on the right and are spread apart on the left. The waves are moving at the same speed in Figures 11.27 and 11.28, so the frequency detected by D will be higher in Figure 11.28 than in Figure 11.27. (More wave cycles pass by the detector during each second when the wavefronts are closer together.) In contrast, the spread-out wavefronts behind the moving source in Figure 11.28 will create a lowered frequency when they are detected by another observer at D'. In both these cases, the detector is stationary; only the source is moving.

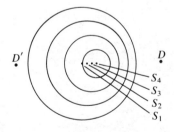

Figure 11.28. Wavefronts of waves emitted by a source that has moved from S_1 to S_4. Compared to the frequency emitted by the source, D detects a higher frequency, and D' detects a lower frequency.

We can discuss the case of a moving detector using either Figure 11.27 or 11.28, but for simplicity, we will use Figure 11.27. When the detector moves toward the stationary source, the detector will pass more wavefronts in the same amount of time than if it were standing still. Thus, the frequency of the detected wave will be higher than if the observer were also stationary. And, of course, the detected frequency will be lowered when the observer is moving away from the source. This effect also occurs for a moving source. That is, for an observer moving toward the source (even when the source also is moving), the frequency will be increased, while an observer moving away sees a lower frequency.

The size of the frequency shift for either a moving source or a moving detector depends on the speed of the source or detector compared to the wave speed. In general, we find a greater source or detector speed results in a greater frequency shift. We should note, however, that there is a fundamental difference between moving sources and moving detectors, even though we usually do not see examples where this difference is important. We have seen that a detector moving away from the source perceives a lowering of the frequency. As the detector's speed increases, the frequency shift becomes greater. However, there is a limit to this effect. When the detector's speed matches that of the wave (or becomes faster), the wave will never reach the detector; nothing will be detected! This phenomenon never happens with a moving source, but something else does. When the source speed is equal to or greater than the wave speed, the cycles of the wave "pile up" in front of the source; the cycles do not have time to get away. This situation leads to new effects, such as sonic booms for sound waves. These new effects are different than the frequency shift of the Doppler effect.

There are several interesting examples of the Doppler shift with sound waves, and there is one interesting example with light waves. We easily can observe the shift in pitch of a sound emitted by a rapidly moving object when the object passes by. Standing next to the track and listening to the horn of a high-speed diesel train is obviously a good example. Auto racing fans also recognize a significant drop in the pitch of the engine's roar as each car changes from "approaching" to "departing." Note that in each of these cases, the Doppler effect is enhanced because the stationary observer starts off in front of and ends up behind the moving sound source. Thus, a frequency higher than that emitted is heard at first, and this frequency changes to one that is lower than the emitted frequency as the source passes.

Since light has a much greater speed than sound, it is difficult to detect any Doppler shift effects for light sources on Earth. The effect is very useful, however, for determining the movement of stars toward or away from us. Amazingly, all stars exhibit the famous "red-shift," which means the color of emitted light is shifted toward the red end of the spectrum, that is, toward lower frequencies. This red-shift means that all stars are moving away from us; we conclude that the universe is expanding.

In summary, the Doppler effect explains how—for any type of wave—the detected frequency is lower (or higher) than that emitted by a source when the source and detector are moving away from (or toward) each other. Either the source or detector (or both) can be moving, and the amount of frequency shift depends on the speed of the moving objects. The explanation fails when the speed of the source or detector becomes as great as the wave speed.

Summary

In order to help us visualize the movement and shape of two- and three-dimensional waves, we developed the concepts of wavefronts and rays. Because wavefronts and rays are always perpendicular to each other, it is easy to construct either a wavefront picture from a ray picture or vice versa. In studying reflection and refraction, we found it useful to construct the wavefront picture first and then develop the ray picture.

Reflection occurs whenever there is an abrupt change in wave speed (due, of course, to a change in the medium). For example, light changes speed as it crosses an interface between air and water, so reflection occurs there. The reflected wave moves back into the same medium through which the incident wave passed. Refraction is a change in direction of wave movement due to a change in wave speed. In contrast to reflection, the refracted wave moves into the medium having the different wave speed. When the wave speed changes slowly, there is a gradual, but continuous change in direction; this situation often occurs with water waves approaching a coast. In contrast, light waves usually make abrupt directional changes as they move from one medium to another; this situation occurs when light enters water or a piece of glass from air.

When a wave is refracted as it crosses an interface, the ray is always closer to the normal on the slower-medium side. The reflected ray, however, makes the same angle with the normal that the incident ray does. Total internal reflection occurs when a wave is moving from a slow medium to a fast one and the incident angle is greater than the critical angle. Light pipes are a practical example of total internal reflection.

We can focus waves with either reflection or refraction at curved interfaces. The behavior of the waves depends on the curvature of the interface and—in the case of refractive focusing—the speeds of the waves on each side of the interface. A widespread use of this effect is with lenses focusing light waves to produce images. A biconvex glass lens in air can produce a real image (that is, one that is projected on a screen). Lenses also can produce virtual images that we can see but cannot project on a screen.

The Doppler effect strongly supports our belief that sound and light are waves. It explains how motion of the source or detector can produce frequency changes. If either the source or detector is moving toward the other, the frequency observed will be higher than that emitted by the source, while relative motion away from each other will produce lowered frequencies. We have difficulty explaining this observation without using waves.

Suggested Readings

The use of light pipes for communication is discussed in two different *Scientific American* articles, "Communication by Optical Fiber" (November 1973) and "Light-Wave Communications" (August 1977). Both of these articles discuss some of the difficulties that occur when we try to send light signals through light pipes, and some of the ideas engineers and scientists have used to solve these problems.

The topics of reflection, refraction, and image formation by lenses are discussed in the section on light in almost every general physics textbook. One such book is *The Dynamic World of Physics,* by Robert T. Dixon (Columbus, Ohio: Charles E. Merrill Publishing Company, 1984).

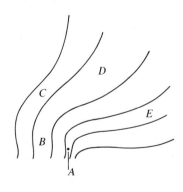

Figure 11.29. For use with review question 1

Review Questions

1. The curved lines in Figure 11.29 show wavefronts for water waves. In which area are the waves moving the fastest?

   a. A

b. B

c. C

d. D

e. E

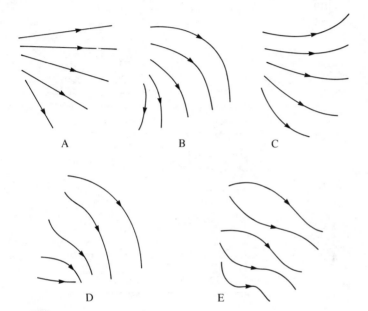

Figure 11.30. For use with review question 2

2. Which set of rays in Figure 11.30 goes with the wavefronts shown in Figure 11.29?

   a. A

   b. B

   c. C

   d. D

   e. E

3. Figure 11.31 shows

   a. rays of waves moving from a slow to a fast medium.

   b. rays of waves moving from a fast to a slow medium.

   c. wavefronts of waves moving from a slow to a fast medium.

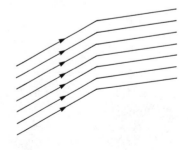

Figure 11.31. For use with review question 3

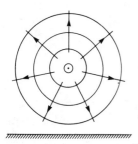

Figure 11.32. For use with review question 4

d. wavefronts of waves moving from a fast to a slow medium.

e. rays of waves being reflected from a plane mirror.

4. Figure 11.32 shows circular wavefronts approaching a plane surface. The reflected wavefronts look like which of the following:

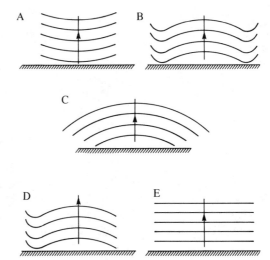

a. A

b. B

c. C

d. D

e. E

Figure 11.33. For use with review questions 5–8

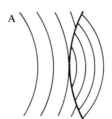
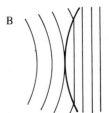

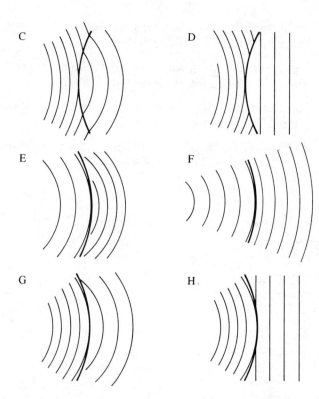

In diagrams A–H, Figure 11.33, circular waves move from left to right across a curved interface. To answer questions 5–8, choose the diagram or diagrams that correctly show the situation described.

5. Waves moving from a slow medium to a fast one across a convex interface.

6. Waves moving from a slow medium to a fast one across a concave interface.

7. Waves moving from a fast medium to a slow one across a convex interface.

8. Waves moving from a fast medium to a slow one across a concave interface.

9. For flint glass, the critical angle is 37°. Thus,

    a. light entering the glass with an incident angle larger than 37° will be totally refracted.

b. light entering the glass with an incident angle larger than 37° will be totally reflected.

c. light leaving the glass with an incident angle larger than 37° will be totally reflected.

d. light leaving the glass with an incident angle smaller than 37° will be totally reflected.

e. light entering the glass with an incident angle smaller than 37° will be totally reflected.

Figure 11.34. For use with review questions 10–13

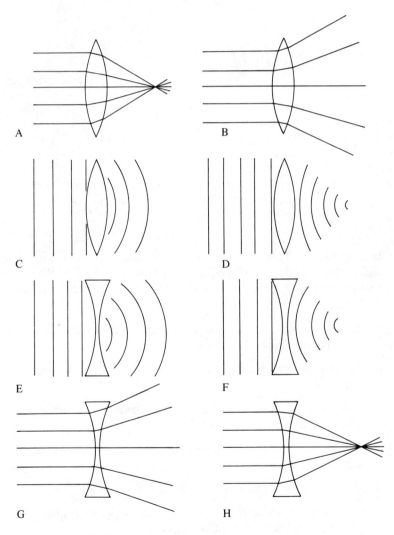

In diagrams A–H, Figure 11.34, wavefronts or rays are shown on both sides of a lens-shaped region. To answer questions 10–

13, choose the diagram or diagrams that correctly show the situation described.

10.   A converging lens.

11.   A diverging lens.

12.   Waves traveling slower through the lens than the surrounding medium.

13.   Waves traveling faster through the lens than the surrounding medium.

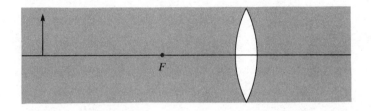

Figure 11.35.   For use with review question 14

14.   Figure 11.35 shows a biconvex air lens under water. Where would the image of the object arrow be located?

a.   Between the object arrow and the focal point, F

b.   Between the focal point, F, and the lens

c.   On the right side of the lens

d.   To the left of the object arrow

e.   There would be no image for this type of lens

15.   On a road, John drives toward Marsha tooting his horn. Marsha immediately notices that

a.   John looks slightly bluer than normal.

b.   John looks slightly redder than normal.

c.   The horn is pitched a little higher than normal.

d.   The horn is pitched a little lower than normal.

e.   The car appears longer than usual.

12

The Eye as a Detector of Light Waves

Important Concepts

- We can compare the eye to a camera. Both have lens systems, an iris, and a light detector (the retina or the film).

- Light is described using the physical variables of wavelength, intensity, and wavelength mixture. These variables are related most closely to the psychological variables of hue, brightness, and saturation, respectively.

- The power of the eye's lens system is adjusted by the lens changing shape. A fatter lens gives greater power so that close objects will be focused on the retina.

- The retina has two types of detectors: rods for vision in low light levels and cones for vision in high levels of light and for color discrimination.

- Color vision is possible because there are three types of cones having response curves with peaks at different wavelengths.

- Defective lens systems in eyes can be corrected with additional lenses. A nearsighted person (who cannot see distant objects clearly) has too strong a lens system (so the image is in front of the retina). A diverging lens is used for correction. In contrast, a farsighted person cannot see nearby objects clearly and needs a converging lens for correction.

We have seen how the energy carried by a light wave can start a chemical reaction on a film emulsion to record the arrival of the light wave. This energy also can initiate a different chemical reaction within our eyes that allows us to see. The human eye is a marvelous instrument. We use it to collect more than 90% of the observations we receive about the world around us. In this chapter, we will approach our study of the eye in a way very similar to the way we studied the ear. That is, we will look at what the normal eye can do, and we will describe the eye in terms of its response to electromagnetic waves. This study also will include a comparison between the subjective and objective ways of describing color. And we will look at the detailed structure of the eye and try and discover how its various parts function. Finally, we will discuss some common vision abnormalities. First, however, we will get an overview of the eye by comparing it to a camera.

Similarities between the Eye and a Camera

It is very useful to compare the eye to a simple camera because both are systems that detect light waves. Thus, they have the same components, although the components may function very differently. Each system has a lens to focus incoming light rays; each has an iris to control the amount of light entering, and each has a light-sensitive substance to record the arrival of light waves. These two systems are shown in Figure 12.1, and their major components are labeled. (The eye will be described in greater detail presently.)

When either system is pointed toward an object, light rays from that object enter the system through a small opening whose size is controlled by the *iris*. When light intensities are low, the iris of the eye changes size automatically, causing the pupil to enlarge so that more light may enter the detecting system. The eye's *pupil* contracts in high intensity light so the system will not be damaged. In a similar way, a photographer changes the aperture of his camera by means of an iris to accomplish the same end, that is, to control the amount of light entering the system. (In some modern cameras, the iris also works automatically.)

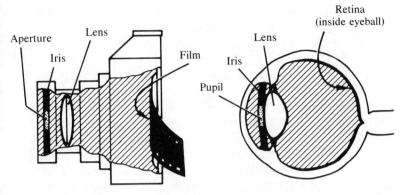

Figure 12.1. Major components of a camera and an eye

When the light rays enter either detector, they are focused by means of a *lens system* onto the actual detecting surface: the film in a camera or the *retina* in an eye. In each case, the image of the object is upside-down and left-right reversed. (When we look at a tree, it appears upright because our brain is trained to interpret the inverted image correctly.) The lens system of the eye will be described in the following material. For the camera, the lens system can range from a simple lens such as the converging glass lens described in the last chapter to much more complicated ones containing many different individual glass lenses or *elements*. In all but the least expensive cameras, the photographer focuses the image onto the film by adjusting the distance between the lens and the film. Recently, cameras that perform this adjustment automatically have been introduced. In either case, this distance must be adjusted because the lens has a fixed shape. In contrast, the distance from the lens to the retina in the eye is fixed, so focusing takes place by changing the shape of the lens. As described in Chapter 10, light striking the camera film initiates a chemical reaction. The same thing happens when light strikes the retina, although, of course, a different chemical reaction takes place. Thus, we see that there are a number of similarities between the eye and the camera as light-detecting systems.

Describing Light

When we discussed sound, we found there were two ways to describe it: with a set of objective (or physical) variables that included frequency, intensity, and vibration recipe or a set of subjective (or psychological) variables that included pitch, loudness, and tone quality. The same type of thing occurs with light.

The objective variables are the same as those used for sound, although we use wavelength instead of frequency and talk about the wavelength mixture rather than the vibration recipe. The subjective variables: *hue, brightness,* and *saturation* are new, however. As with the case of sound, the two sets of variables for light are related.

When there is only one wavelength of light, this relation is particularly simple. (The same thing happens for sound: The pitch of a pure tone depends only on its frequency.) The color—or more precisely, the hue—of light having only one wavelength (or frequency) is determined by that wavelength. Such a light is termed *monochromatic* (meaning one color) or *spectrally pure.* When light contains many wavelengths, the relation between the variables is more complicated. Figure 12.2 shows the primary relationships between these two sets of variables for light. As mentioned, the basic subjective term we use to describe the color we are observing is hue. And hue is determined by the dominant wavelength of the light we are looking at. (Dominant wavelength in this case means the wavelength that has the largest amplitude). The word brightness is analogous to loudness for sound in that brightness is primarily determined by the amplitude or intensity of the light we are observing. However, as we will see when we discuss the response of the eye's color receptors, the brightness also can be influenced by the wavelength of light. (Note the similarity of this relationship to the way in which the loudness of a sound partly depends on the sound's frequency.)

Saturation is related to the mix of wavelengths in the light and is thus analogous to the tonal quality of a sound. When a light is fully saturated (that is, its saturation is 100%), then we have monochromatic light. In contrast, a light that is a mixture of all frequencies (white light) is completely unsaturated. To illustrate the term saturation, we can describe what happens when we start with one hue and change its saturation. In this example, we start with an additive blue primary. This color appears blue, so that is its hue. To *increase* this light's saturation, we pass the light through filters to eliminate some of the

Figure 12.2. Terms used to describe light

Objective (physical)		Subjective (psychological)
Wavelength	←——→	Hue
Intensity (amplitude)	←——→	Brightness
Mixture of wavelengths	←——→	Saturation

frequencies that are included in the blue end of the spectrum. By being careful, we can eliminate frequencies of light in such a way that the hue does not change. However, the more frequencies we eliminate, the purer or more saturated the light becomes. We could have gone in the other direction, however, by adding small amounts of white light to the original primary blue. Since white light contains all frequencies, we would be decreasing the saturation of the original blue light. As we add more and more white light, of course, the blue will become paler. At some point, we would change the description of the hue from pale blue to bluish white.

The hue of a color is analogous to the pitch of a sound in that both are determined by the frequency (or wavelength) of the respective wave. As children, we learn our colors, and when shown a light having a wavelength of 580 nm, we identify it as yellow. But as Thomas Young first discovered in 1802, we also will identify a particular mixture of monochromatic red and green lights as yellow. Young originated the *tricolor* theory of vision, which states that any spectral color can be duplicated by adding the correct intensities of monochromatic lights in the red, green, and blue portions of the spectrum. Although his idea was forgotten for fifty years, it was rediscovered by James Clerk Maxwell and Herman Von Helmholtz and is currently the accepted theory of color vision.

Many different systems have been invented to organize the identification of color. The most scientific of these systems is the CIE system, first developed in 1931. (CIE stands for Commission

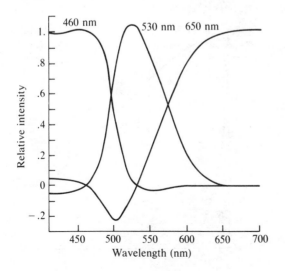

Figure 12.3. The intensitites of three primaries needed to match the pure spectral colors along the bottom

Internationale de l'Eclairage, a French-based international group.) This system uses primaries with specific wavelengths in the blue, green, and red parts of the spectrum. A large number of people were tested, and the "standard" curves of Figure 12.3 were developed. These colors show the relative intensities of each primary color needed by an average person to match a monochromatic sample of the given wavelength. Using these values as coordinates, each color becomes a specific point on a *chromaticity diagram*. (See Figure 12.4 and Plate 5 in the color section.) On this diagram, each color in the visible spectrum appears around the curved edge; these colors are 100% saturated. The magentas (or purples) that do not appear in the spectrum (because they are mixtures of blue and red) also are located on this diagram; they are along the straight line connecting the blue with the red.

We described earlier two sets of primary colors: the additive primaries of red, green, and blue, and the subtractive primaries of cyan, yellow, and magenta. These two sets of primaries are used by physicists, but other groups use different sets of primary colors. For example, psychologists helped to develop the CIE color system just described. They also have experimented with the use of four primary colors: red, yellow, green, and blue. Even more familiar are the artist's primaries of red, yellow, and blue. This set is similar to the subtractive primary colors because artists' pigments act very much like filters. For example, a pigment is red because all colors except red are absorbed by the pigment when white light falls on it. This reaction is just what happens with a red filter, except the red light passes through the filter;

Figure 12.4. Chromaticity diagram showing how colors can be described using two coordinates (*x* and *y*). The numbers around the edges are pure spectral colors. Mixtures of pure colors are inside the figure.

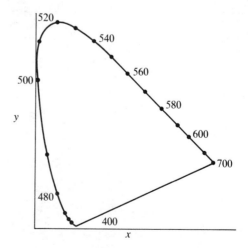

whereas, it is reflected from the pigment. But in each case, all other colors are absorbed.

Each of these various sets of primary colors is useful in particular applications. Their usefulness lies in their ability to make the eye believe it is seeing some color of the spectrum when it really is seeing a mixture of the primary colors.

Capability of the Eye

In describing how well the normal eye works, we must consider its response to light both in terms of frequency and intensity and the size of the objects it can detect. This task is complicated by the fact that there are two separate visual systems in the eye, one for daytime use (that is, when there are high intensities of light) and the other for nighttime use (when the intensity of light is very low). This visual duality occurs because there are two different types of light receptors in the eye. These receptors will be described when we discuss the structure of the eye.

The eye's night vision is much more sensitive to light than its day vision, as illustrated in Figure 12.5. This figure is, of course, another example of a response curve. It shows that both of the eye's visual systems have broad resonances near the middle of the range of frequencies included in visible light. (The eye does not respond at all to any frequency outside this range; this situation explains why such radiation is invisible!) We call the higher response curve (or sensitivity) of one of the eye systems night vision because we can use it to see when our day vision is

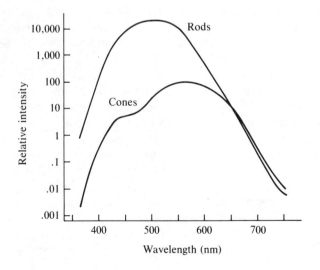

Figure 12.5. Response curves for rods and cones (adapted from G. Wald, *Science 101*, 653–58 (1945), Copyright 1945 by the American Association for the Advancement of Science.)

not responding. But we pay a price for this higher sensitivity. Night vision has no ability to discriminate between various frequencies; that is, we cannot determine color in dim light. This fact is very easy to check. In the evening when the last light of day is fading, the flowers in a garden retain their shapes but not their colors. The flowers all appear to be various shades of gray. Some flowers may look brighter than others and, thereby, may be easier to see, but they are no longer colored. Another interesting feature of our night vision is that we can see objects better if we do not look directly at them. Astronomers, for example, can see fainter stars through their telescopes by looking off to one side of rather than directly at the star.

During daylight illumination (or with bright sources), we can see the full range of spectral colors. We also can visually examine an object best when we look directly at it. Both of these aspects of day vision are in direct contrast to night vision as described above. The question of how the eye perceives color is still not fully understood. As mentioned previously, the currently accepted ideas on this subject started with the experiments of Thomas Young around 1800. On the basis of his tricolor theory and later confirming work, scientists developed the idea that there are three types of color receptors for day vision. As one looks at a color, these receptors are stimulated in just the right way to tell the brain what color is being observed. When we look at the eye's structure, we will describe this theory of color vision.

But there are some colors that cannot be duplicated with three primary colors. No matter how we mix the additive primaries, we cannot produce brown, and brown does not appear anywhere in the rainbow. Thus, there must be more to color vision than just having three types of color receptors. One of the reasons for these difficulties in understanding color perception is that all references to color are subjective. Each individual interprets the frequencies of the light he sees in his own way. Tests have been devised that allow scientists to compare the ways different people respond to various mixtures of colors. But we certainly do not understand fully how color vision works.

As in our discussion of the ear's capability, I include here a brief description of how well the normal eye can discriminate between colors and brightnesses. (This ability for the eye is analogous to differential pitch discrimination and differential loudness discrimination for the ear.) For *differential hue discrimination,* we find that the eye can distinghish a difference between two pure colors whose wavelengths differ by about 1–7 nm.

Smaller differences can be distinguished in the yellow and blue-green parts of the spectrum, while larger separation between wavelengths is needed for color detection at the two ends of the visible spectrum. The three-color theory of visual perception and the structure of the retina deduced from that theory help us understand why the eye has greater hue differential discrimination in the yellow and blue-green parts of the spectrum. Since a knowledge of the eye's structure aids in understanding this effect, we will postpone the explanation until after we have discussed the structure of the retina. For now, we note that hue discrimination does not follow Weber's law (which we mentioned in connection with the ear's discrimination capabilities).

In contrast, we find that for high levels of illumination, *differential brightness discrimination* does obey Weber's law approximately. That is, the change in brightness that we can detect divided by the brightness itself is almost constant—about 5%. Weber's law is not valid for low levels of illumination because we need much larger changes to detect a difference. Thus, we find that the ability to detect changes in intensities is similar for sound and light in that both the eye and ear approximate Weber's law for high intensity levels but require larger changes for low intensities.

The Structure of the Eye

Now, we will examine the structure of the eye in greater detail. A more complete cross-sectional view of the eye is shown in Figure 12.6. Light entering the eye crosses the *cornea* and passes through the *aqueous humor,* the *lens,* and the *vitreous humor* before it reaches the *retina*. As the light passes through each of these substances, it changes speed, so refraction occurs at each interface. The greatest change in speed occurs when the light passes from air through the cornea into the aqueous humor. Thus, the greatest refraction occurs at the surface of the cornea. The lens, on the other hand, is used to make only small adjustments in the refractive power of the total lens system. These adjustments are accomplished by the lens changing shape. When the eye focuses on nearby objects, the lens gets fatter as shown in Figure 12.7A. This fattening increases the curvature of the lens' surfaces so that the light rays are refracted through the required greater angles. When distant objects are being observed, not as much refracting power is needed (the light rays do not change direction as much), so the lens becomes thinner and flatter as shown in Figure 12.7B. The aqueous and vitreous

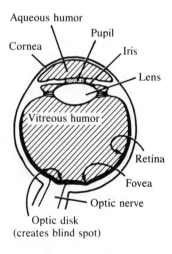

Figure 12.6. A cross-sectional view of the eye

Figure 12.7. The shape of the eye's lens for observing close and distant objects

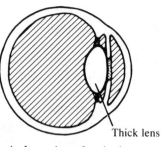
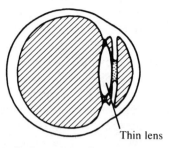

Thick lens

Thin lens

A. Lens shape for viewing a close object

B. Lens shape for viewing a distant object

humors are gelatine-like substances. They help to maintain the shape of the eyeball and the positions of its components without interfering with the light rays passing through.

The retina contains the light-sensitive endings of nerves. When light strikes these nerve endings, impulses are sent to the brain. By unscrambling these nerve signals, the brain interprets the image on the retina. Two types of nerve endings occur in the retina; they are called *cones* and *rods* because of their respective shapes. The rods give us night vision, and we cannot discriminate among colors with rods because each rod in the retina has the same response curve over the entire visible part of the spectrum. The rods are more sensitive to light, however, and thus, we use them for our night vision. While the cones are less sensitive to light (so higher light intensities are needed before they will send nerve impulses to the brain), there are three types of cones with different response curves, so the cones are able to differentiate among colors.

The response curves for cones are shown in Figure 12.8. Because the curves peak in the blue, green, or red portions of the visible spectrum, the cones are called blue, green, or red cones, respectively. To see how color vision works, let's consider an example. Suppose we look at a pure green light of wavelength 600 nm (indicated by the dashed vertical line in Figure 12.8). This light wave will cause nerve impulses to be sent by only red and green cones. When the impulses from these types of cones arrive, the brain sorts them out and determines that the color is yellow. Note that if there were only green cones in the retina, then there would be no way to tell the difference between the yellow light of wavelength 600 nm and the blue-green light of wavelength 500 nm. The green cones respond equally to both colors. Also, we can see why the eye can be fooled into thinking that a particular mixture of red and green lights is yellow. If a red light ($\lambda = 700$ nm) of the proper intensity is combined with a green light (λ = 550 nm), then the relative stimulation of the red and green cones can be made the same as that occurring for the yellow light.

We now can see why the eye can discriminate so well between different hues in the yellow and blue-green parts of the spectrum. In each of these regions, there is a cone that has a rising response curve and another cone that has a falling response curve. Thus, in these regions of the spectrum, a small change in a light's wavelength produces a larger response for one type of cone and a smaller response from the other. When an increase is combined with a decrease in this way, our ability to detect the change is enhanced.

Rods and cones are not uniformly distributed across the retina. There is a very high concentration of cones in the *fovea,* a small region of the retina directly behind the lens system. We use this region of the retina when we look directly at an object. As you read this page, you move your eyes so that the word you actually are reading is focused on the fovea. The higher concentration of cones in the fovea means that we have greater powers of resolution for day vision than for night vision, but this concentration also prevents any rods from being in the fovea. Thus, when we are trying to detect something with our night vision, we actually can see the object better if we do not look directly at it. By looking to one side of the object, we focus its image on the part of the retina that is away from the fovea and has, therefore, some rods to detect the light.

There is another interesting area of the retina that causes the *blind spot.* At this location, all of the nerves leading from the

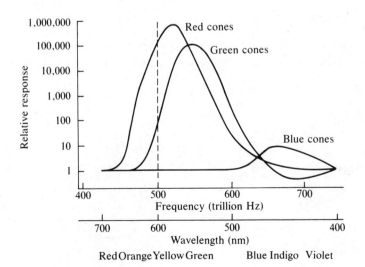

Figure 12.8. Response curves for cones

Figure 12.9. Layout for observing the blind spot

rods and cones distributed over the retina pass through the retina on their way to the brain. This bundle of nerve cells is called the *optic nerve.* As it passes through the retina, it leaves no room for nerve endings, so there is a blind spot in the eye's field of vision. This blind spot occurs for both our day and night vision. Note that the optic nerve passes through the retina on that side of the eyeball toward the center of the head. This place of passage means vision is lost toward the outside of the field of vision.

We can see the effects of this blind spot with an experiment that uses Figure 12.9. To perform this experiment, hold the page at arm's length, close your left eye, and look directly at the *X* with your right eye. As you look at the *X,* you should be able to see both the circle and the square "out of the corner" of your eye. Now, keep your eye pointing at the *X,* and move the page slowly toward you. When it is about a foot from you, the square will disappear. As you continue to move the page closer to your face, the circle will disappear, and the square will reappear. At even closer distances, the circle also will reappear. When each object disappeared, its image was being focused on the part of the retina that causes the blind spot, as indicated in Figure 12.10. The experiment also can be done with the left eye, but the book must be turned over so that the *X* is on the inside.

Eye Defects

Here we will discuss defects of the eye that occur for basically two different reasons: problems with the lens system and prob-

Figure 12.10. Focusing of an object on the part of the retina that causes the blind spot

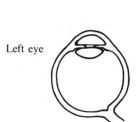

Left eye

Right eye

A. Distant object   B. Close object

Figure 12.11. Focusing of light rays for a farsighted eye

lems with the nerves whose endings are embedded in the retina. When the lens system of our eye works improperly, we cannot bring all objects into sharp focus on the retina, which means the image is blurred so the object we are looking at appears blurred. People who can focus distant but not nearby objects well are said to have *farsightedness* or *hyperopia*. The opposite condition, termed *nearsightedness* or *myopia* is when a person can focus close objects onto the retina but not distant ones. A third type of defect is called *astigmatism*. A person with this defect will have some parts of an object in focus and other parts out of focus at the same time.

Figure 12.11 shows how the light rays from a distant and a nearby object converge to form an image in a farsighted eye. The evidence for this defect shows up when looking at a close object. Since the refractive power of the lens system is too low, the light rays are not bent enough when viewing the nearby object. In this situation, the image would be in focus behind the retina, meaning the image on the retina is out of focus. The same two situations are shown in Figure 12.12 for a nearsighted eye. In this case, the refractive power is too great, so distant objects are focused at a point in front of the retina, and the image on the retina is blurred. Both of these conditions can be corrected by using an extra lens. The extra lens can be in a pair of glasses or can be a contact lens. To correct farsightedness where the refractive power is too small, we use a converging lens. A diverging lens is used to correct nearsightedness because the refractive power of the eye's lens system must be reduced. These two situations are illustrated in Figure 12.13.

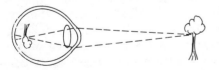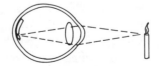

Figure 12.12. Focusing of light rays for a nearsighted eye

A. Distant object   B. Close object

While everyone's color vision is not exactly the same, certain differences from the average are so great that we term the vision defective. We refer to this type of defect as *color blindness;* many specific forms of color blindness occur, but in only one of them is a person truly color blind. This condition, termed *monochromatism* (meaning "one color"), is extremely rare. About one person in every 30,000 has this abnormality. A true monochromat sees the world only in shades of gray, as if he always were watching a black and white television set. A possible explanation for this condition is that the retina contains no cone-type receptors. Since the rods have no ability to detect color, total color blindness results.

Dichromatism is a much more prevalent form of color vision disability; it occurs in about 2% of white males, although only 0.03% of females are afflicted. A dichromat can match any of the colors in the spectrum using only two colors rather than the three required by persons with normal vision. Thus, only two primary colors will produce all spectral colors for a dichromat. The usual interpretation is that a person suffering from dichromatism has only two types of cones. A common form of dichromatism is red-green color blindness, where a subject cannot distinguish between reds and greens. Since many of these people have some limited discrimination between these colors, it also seems possible that the response curves for the red and green cones lie much closer together than for people with normal color vision. As a result, full differentiation is impossible.

Another type of color blindness can be explained in the same way. *Anomalous trichromatism* is the term used to describe the condition where someone needs three primary colors to match all the spectral hues, but uses amounts of each primary that are different from a person with normal vision. This condition is easy to explain by shifting the response curves of the different types of cones. This explanation, however, does not tell us why the response curves have been shifted!

In summary, we find the eye is an extremely effective detector of light. No photographic film has been developed yet that can match the eye's ability. In making this statement, however, we must remember the differences between these two types of detectors. The eye's record of what it sees is transient. When an image forms on the retina, it is interpreted by the brain and then is gone. (The eye does have *persistence of vision,* which means that the image does remain for a fraction of a second, but no longer). Images on film are permanent, of course, but the record on the film is also cumulative; the effects of successive exposures

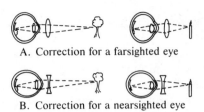

A. Correction for a farsighted eye

B. Correction for a nearsighted eye

Figure 12.13. The use of eyeglasses to correct visual defects

add. Thus, film can be used to record very low intensities of light that last over some length of time. But for observing the world around us, our eyes are superb!

Summary

Comparing the eye to a camera is useful because both systems have many common functional parts. Each has a lens system to focus the light waves and an iris to limit the amount of entering light. The retina in the eye is similar to the film in a camera because it records the arrival of incident light. In each, the record is produced because the light initiates a chemical reaction.

We describe light physically by using the variables of wavelength, intensity, and wavelength mixture. These variables are most closely related to the respective psychological variables of hue, brightness, and saturation. Of the many schemes used to classify colors, the CIE chromaticity diagram is the one in widest use. It was obtained by testing the responses of a large number of people to various mixtures of pure colors.

The retina contains two types of receptors: rods and cones. Rods are useful for vision under low levels of illumination; rods are up to one hundred times more sensitive to light than cones. Although cones are less sensitive than rods, cones have the ability to detect colors. The three-color theory of color vision subscribes to the idea that three types of cones exist: red, green, and blue. Each of these types has a different response curve. Thus, a given monochromatic light produces certain reactions in each type of cone that the brain interprets as a specific color.

The lens system of the eye consists of several elements. Most of the refraction occurs at the surface of the cornea, where the greatest change in light speed occurs. The lens, itself, is used to adjust the refractive power of the system so that both near and distant objects can be focused on the retina. When the system is defective, additional lenses are used for correction. For example, a converging lens corrects a farsighted eye (which can focus distant objects correctly, but which creates an image of close objects behind the retina).

A person who is "color blind" has some defect in his color-sensing apparatus. The most serious form of color blindness is monochromatism, for which all objects appear as various shades of one color. People having dichromatism can match all colors of the spectrum with mixtures of only two monochromatic sources. Anomalous trichromats require three color sources, but mix these sources in ways that are different from ways used by people with normal color vision.

Suggested Readings

The complicated process of seeing is discussed nicely in two different books. At a lower level, *Light and Vision,* by Conrad G. Mueller, Mae Rudolph, and the editors of *Life* (New York: Time Incorporated, 1966) does a very nice job. There are a large number of excellent photographs and many topics that we have not covered, such as visual illusions. At a slightly more advanced level is *Eye and Brain, The Psychology of Seeing,* by R. L. Gregory (New York: World University Library, McGraw-Hill Book Company, 1966). The book also covers a number of topics not covered in this book.

The comparison between the two light-detecting systems is carried a bit further than we have done in the article "Eye and Camera" (*Scientific American,* August 1950). The comparisons we have made and additional ones involving the chemistry of the retina and film are made.

Edwin H. Land (the founder of the Polaroid Corporation) has discovered some interesting aspects of the way we see that do not fit into the tri-color theory of color vision. His ideas are described in two articles in *Scientific American:* "Experiments in Color Vision" (May 1959) and "The Retinex Theory of Color Vision" (December 1977). In the first article, he describes how two beams of light (of different wavelengths) projected through black and white transparencies onto the same screen will produce a fully colored scene. The second article gives his interpretation of what this experiment means for color vision. A college student's experiments to duplicate Land's results are discussed in "The Amateur Scientist" column in the June 1979 issue of *Scientific American.*

Review Questions

1. In comparing the eye with a camera, we find they are different because

a.   only the eye has an iris.

b.   only the camera has an enclosure.

c.   the eye has persistence of vision, while the camera has a permanent image.

d.   the eye has an iris, while the camera has an aperture.

e.   light initiates a chemical reaction on the film in a camera but causes the rods and cones to vibrate back and forth in the eye.

2.   The psychological variables of hue and saturation used to describe light are most closely related to which physical variables?

a.   Wavelength and intensity

b.   Color and brightness

c.   Intensity and wavelength mixture

d.   Wavelength and wavelength mixture

e.   Frequency and brightness

3.   The chromaticity diagram shows

a.   the relative response of each type of cone to various wavelengths of light.

b.   a comparison of how rods and cones react to light.

c.   a way of plotting all colors in terms of two variables.

d.   all the colors that are complementary pairs.

e.   which three colors are the additive primaries.

4.   The greatest amount of refraction in the lens system of the eye occurs when the light

a.   passes through the middle of the lens.

b.   passes from the vitreous humor to the aqueous humor.

c.   passes from air into the vitreous humor.

d.   enters the cornea.

5.   The old saying "at night, all cats appear gray" has the following scientific basis:

a.   At low light levels, all cones are stimulated, so every object appears white or off-white.

    b. At low light levels, the cones are not stimulated, and rods cannot distinguish colors.

    c. At night, only the fovea reacts to light.

    d. In low light levels, the lens transmits only blue light, which we interpret as gray.

    e. At night, the optic nerve only reacts to light and dark.

6. When we view a magenta light,

    a. blue and red cones in the retina are stimulated.

    b. blue and green cones in the retina are stimulated.

    c. red and green cones in the retina are stimulated.

    d. magenta cones in the retina are stimulated.

    e. since green is the complementary color to magenta, only green cones in the retina are stimulated.

7. A red-green dichromat

    a. sees all objects as shades of red and green.

    b. cannot distinguish between red and green.

    c. has lost all red and green vision and sees the world in shades of blue.

    d. is nearsighted for red light and farsighted for green light, or vice versa.

    e. reverses reds and greens.

8. A farsighted person

    a. can see distant objects clearly and needs a converging lens for correction.

    b. can see distant objects clearly and needs a diverging lens for correction.

    c. cannot see distant objects clearly and needs a converging lens for correction.

    d. cannot see distant objects clearly and needs a diverging lens for correction.

13

Other Wave Behavior

Important Concepts

- Diffraction is the bending of a wave around a barrier or the spreading of a wave passing through an opening. It can occur for all types of waves.

- Longer wavelength waves exhibit more diffraction than short wavelength ones.

- A diffracted wave is split into different beams; between each pair of adjacent beams, there is a diffraction minimum, which is a region having no wave.

- Interference occurs when waves from two or more coherent sources superpose. We have constructive interference where the waves are in phase, so they add to produce larger waves. At points where the waves are in antiphase, they tend to destroy each other, and we have destructive interference.

- Interference can occur with all types of waves. The interference pattern depends on the distance between the sources, compared to the wavelength and on the phase between the sources.

- Interference of light waves occurs both for the case of two slits illuminated by a single source and for the case of light reflected from a thin film.

- Scattering is the incoherent reflection of light from many small objects. There are two types of scattering: Wavelength-independent scattering occurs when waves are scattered from objects larger than the wavelengths of the waves, while objects smaller than the wavelengths involved produce wavelength-dependent scattering. In the latter case, shorter wavelength waves are scattered better than long wavelength waves.

- A wave is polarized when all its vibrations are in the same direction. Only transverse waves can be polarized; since we can polarize light, it is a transverse wave.

We have seen how an understanding of wave behavior can help explain some of the things we observe in the world around us. Optical instruments (including the eye) work because of the refraction of light waves. White light passing through a prism is broken into its constituent colors because of dispersion. An organ pipe sounds a note of a certain pitch because standing waves of only certain frequencies occur within the pipe. There are other events in the world around us that we can explain using waves, but in order to do so, we need to first investigate other ways in which waves behave. The specific new behaviors we will be looking at in this chapter are diffraction, interference, scattering, and polarization. As before, we will develop each of these ideas using waves we can observe directly. Then, we will extend the idea to other types of waves.

Diffraction

The easiest place to observe the diffraction of waves is when a wave passes through an opening that is about the same size as the wavelength of the wave. For example, we can see water waves diffracting when they pass through an opening in a barrier. We can set up this situation in a ripple tank, or we might observe the same effect for ocean waves passing through an opening in a breakwater or a narrow opening to a bay. In such a case, when plane waves having a wavelength about equal to the size of the opening pass through the opening, they spread out in

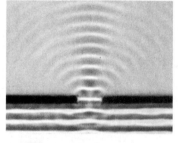

The diffraction of water waves moving through a narrow opening. Note the regions of calm water between zeroth- and first-order beams. (PSSC Physics, 2nd edition, 1965; D. C. Heath and Company with Education Development Center, Inc. Newton, MA)

many directions. This spreading of the waves, or more precisely, the bending of the waves around the barrier, is what we call *diffraction*. The amount of the diffractive spreading changes as the ratio of the wavelength to the opening size changes, and the spreading is also not uniform.

Figure 13.1 shows the wavefronts of some plane waves being diffracted as they pass through an opening. As we can see, the diffracted wave is broken up into several distinct beams of waves. The beam that contains that part of the wave moving straight ahead is called the *zeroth-order beam*. On each side of this zeroth order beam, higher order beams form. The first one, of course, is the *first-order beam,* and the others follow in turn.

The change in the spreading of the diffracted wave as the ratio of the wavelength to opening size changes is illustrated in Figures 13.2 and 13.3. In Figure 13.2, the wavelength is varied from picture to picture with a fixed opening size. Picture A shows the case where the wavelength of the wave is much smaller than the width of the opening. Thus, very little diffraction occurs. In pictures B through E, the size of the opening is kept the same as in picture A, but the wavelength of the wave is increased. As the ratio of the wavelength to the opening size becomes larger, all the diffracted beams spread apart and become wider. Figure 13.3 is similar to Figure 13.2, except that instead of the wavelength changing from picture to picture, the opening size changes. This figure is included to emphasize the fact that it is the ratio of the wavelength to the opening size that controls the amount of spreading and number of higher order beams that occur when a wave is diffracted. Notice that in all of these figures, there are distinct regions between the diffracted beams where waves do not occur. Indeed, there is a point between each pair of diffracted beams where the amplitude of the waves becomes zero. These regions are called *diffraction minima.*

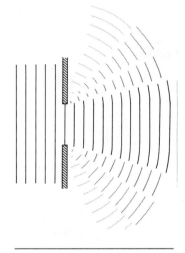

Figure 13.1. The diffraction of waves passing through a narrow opening in a barrier

The change of diffraction through a narrow opening as the wavelength changes. Note the increased diffraction for larger values of λ/w due to the larger wavelength. (PSSC Physics, 2nd edition, 1965; D. C. Heath and Company with Education Development Center, Inc. Newton, MA)

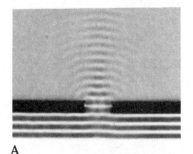

A

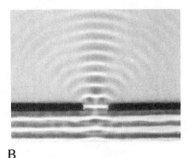

B

C

Figure 13.2. The change in diffraction for different wavelength waves passing through a 1 cm opening

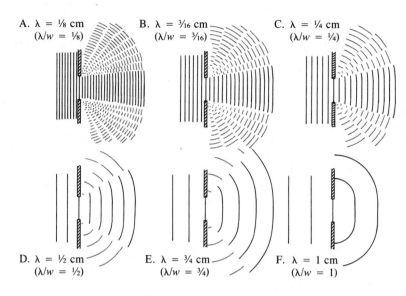

A. λ = ⅛ cm
(λ/w = ⅛)

B. λ = ³⁄₁₆ cm
(λ/w = ³⁄₁₆)

C. λ = ¼ cm
(λ/w = ¼)

D. λ = ½ cm
(λ/w = ½)

E. λ = ¾ cm
(λ/w = ¾)

F. λ = 1 cm
(λ/w = 1)

Figure 13.3. The change in diffraction for ¼ cm wavelength waves passing through different sized openings

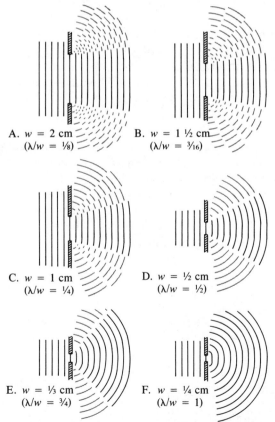

A. w = 2 cm
(λ/w = ⅛)

B. w = 1 ½ cm
(λ/w = ³⁄₁₆)

C. w = 1 cm
(λ/w = ¼)

D. w = ½ cm
(λ/w = ½)

E. w = ⅓ cm
(λ/w = ¾)

F. w = ¼ cm
(λ/w = 1)

Diffraction also occurs around the edge of a barrier. This diffraction is much less obvious, so we have to look more carefully for its effects. We can make this observation using waves in a ripple tank by increasing the size of the opening in a barrier until the opening is much larger than the wavelength of the waves. As the opening size is increased, we must watch the different orders of the diffracted beams carefully as they move toward the forward direction. Not all of the orders disappear, but they do become weaker, so they are harder to see. When the opening size becomes quite large compared to the wavelength of the wave, a further increase in opening size has no effect on the diffracted wave. Thus, we concentrate our attention on the wave being diffracted around the barrier on one side of the opening and move the barrier on the other side of the opening. When the opening is large, the movable barrier can be removed entirely without affecting the wave that is diffracting around the stationary barrier we are watching. We now are observing the diffraction of these waves around the edge of an obstacle, rather than the diffraction of waves through an opening.

One way of understanding why it is easier to observe diffraction when the wave passes through a small opening than when it passes by an obstacle is to recognize that the edge of an obstacle affects the wave in both the region that is in the "shadow" of the obstacle and in the region that is not in the shadow of the obstacle. Since an opening is just two obstacles, if they are close together, each obstacle will affect the entire wave behind the opening. The fact that both sources of diffraction contribute enhances the overall effect and makes it easier to see.

Although the diffraction of waves passing by a single obstacle is not as obvious as that for waves passing through an opening, the effect is definitely there. Just as for waves passing through an opening, there is a greater amount of diffraction for long wavelength waves than there is for short wavelength ones. Another way of saying this statement is that long wavelength waves bend more when passing an obstacle than short wavelength ones.

Now, we will complete our discussion of diffraction by looking at the observable effects of it in sound and light. When we compared audible sound waves with visible light waves, we noted that the sound waves are much longer. Thus, we fully expect that sound waves are diffracted to a much greater extent than light waves. This conclusion is indeed the case; we can hear around corners that we cannot see around. Of course, the low-pitched (long wavelength) sounds are diffracted more easily than the high-pitched sounds. But even the high-pitched sounds are diffracted much more effectively than any visible light waves. (The wavelength of the highest pitched audible sound is much longer than that of any visible light wave.)

As we consider the diffraction of light waves, we initially are interested in the diffraction of white light. Recalling that white light is a mixture of all wavelengths of visible light, we recognize that the different components of the white light will be diffracted to different degrees. Red light will be diffracted the most, while blue light will be diffracted the least. Note that this effect is the opposite of the effect that occurs when light is dispersed during refraction; there, blue light is bent the most, and red light is bent the least. This difference in behavior will allow us to determine which effect is occurring when we see white light being broken up into its constituent colors by one of these two processes.

Since the wavelength of any visible light is very short, special conditions must be used to observe diffracted light waves. Just as with water waves, it is easier to see the effects of diffraction of light waves when the waves pass through a small opening. Significant effects occur when the size of the opening is about the same as the wavelength of the wave. This situation means that for light, we must have a very small opening. One easy way to obtain such a small opening is to use a small piece of heavy paper such as an index card. With a razor blade or a sharp knife, we can make a single cut about two inches long through the card. The position of such a cut is shown in Figure 13.4. Now, by pressing down on the card on one side of the cut, such as at point A in Figure 13.4, and by lifting up on the other side of the cut, such as at point B in Figure 13.4, we can change the width of the opening created by the cut. Using this technique to create a narrow slit and looking at a small light source (such as a flashlight bulb across the room or a distant street light), we can see the effects of diffracted light very easily. We can change the width of this opening by holding point B fixed and by pushing harder or softer on point A. As the width of the slit is changed, the pattern of the light that we see also changes. For a very narrow slit, the light is spread out in a direction perpendicular to the slit.

As we increase the width of the slit, several dark spots will appear in the spread-out light and will move toward the slit as the slit gets wider. These dark spots are the diffraction minima that we observed with water waves. The various diffracted beams between these minima can be identified in the same way they were for diffracted water waves. The zeroth order beam is

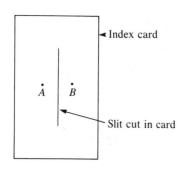

Figure 13.4. Using an index card to view diffraction of light

the one that is in the direction of the slit itself. Then, on either side of this are the two first order diffraction beams, and then the second ones, and so forth.

If we look very carefully at the edges of the diffracted beams (that is, next to the dark spots), we may be able to see a slight coloring of the white diffracted beams. This coloration occurs because of the wavelength-dependent effect of diffraction. Red appears near the outer edges of each diffracted beam, and blue appears near the inner edges. The color appears only next to the edges of the white diffracted beams because these beams are composed of the diffracted beams of each color. Thus, the central part of each diffracted beam contains all the colors and appears white. At the outer edge of each beam, however, the blue end of the spectrum does not contribute because its diffraction minimum occurs there. In the same way, the inner side of each diffracted beam does not contain the red end of the spectrum. (Remember, red light is diffracted more than blue light.)

There are other materials we can use to observe the effects of light wave diffraction. A piece of fine-mesh wire screen, for example, has narrow openings in both directions. Since diffraction makes the light spread out in a direction perpendicular to the opening, light passing through this screen will be spread out in two directions that are perpendicular to each other. The regular pattern of the fibers in a piece of cloth such as silk or nylon also will diffract the light in two perpendicular directions.

Just as for other waves, it is more difficult to observe the diffraction of light passing by the edge of an obstacle as compared to light passing through a narrow opening. To make this observation easier, we should use one color of light so that we eliminate the overlapping of the diffracted beams of different colors that occurs with white light. It would be even better, of course, to use light of a single wavelength. As we will see in Chapter 15, a laser is a device that produces light of a single wavelength, and so is suited ideally for this purpose. If we look at the edge of the shadow of an object illuminated by one wavelength of light, we see a series of bright and dark lines parallel to the object's surface. It is impossible to determine the location of where the object's shadow should start by observation. This effect is illustrated in Figure 13.5.

Interference

Interference is another effect that can be explained using the principle of superposition. This effect occurs when waves are

Figure 13.5. The shadow of a razor blade showing diffraction around the edges (Sears, Zemansky & Young, UNIVERSITY PHYSICS, Sixth Edition, © 1982, Addison-Wesley, Reading, Massachusetts. Pg. 788, Fig. 41–16. Reprinted with permission.)

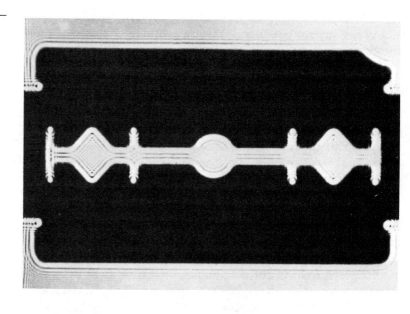

The interference pattern of waves generated by two sources oscillating in phase. The sources are about two wavelengths apart. (PSSC Physics, 2nd edition, 1965; D. C. Heath and Company with Education Development Center, Inc. Newton, MA)

emitted in generally the same direction from two or more separate sources. These sources can be original sources of the waves, or they can be openings in a barrier that allow the waves to pass through and thereby act as separate sources of waves beyond the barrier. As the waves move outward from these sources into the same region, the waves add together according to the principle of superposition. As in the previous cases when we used the principle of superposition, at some places, the waves will add together to produce bigger waves, and at other places, the waves will combine to produce smaller waves. We say that *constructive interference* is occurring when bigger waves are produced and that there is *destructive interference* when smaller waves are produced. We first will study the patterns of constructive and destructive interference when water waves are coming from two sources. Then, we will look at the same effect for both light and sound.

A ripple tank is ideally suited for observing the interference of two sets of water waves. Let's consider two objects (bobbers) that we push up and down at the surface of the water. As these bobbers move up and down, they produce waves that move outward from the bobbers with circular wavefronts. Figure 13.6 shows (at a particular instant of time) the location of a series of wavefronts that were generated by each of two bobbers. The solid lines are wavefronts drawn along the crests of successive waves, and the dashed lines are wavefronts drawn along the troughs. We see that at point P, there is both a crest from bobber

A and a crest from bobber B. Since both these crests are located at point P at this time, they add to produce a large crest at point P. As the waves continue moving outward, the trough behind the crest from bobber A will reach point P at the same time that the trough behind the crest from bobber B does. (Both waves are traveling at the same speed so they move the same distance in the same time.) They will add to create a much deeper trough than either of the troughs individually. Thus, at point P, we get higher crests and deeper troughs; we have a point of constructive interference. In contrast, destructive interference occurs at point Q because a crest from one of the bobbers is always arriving there at the same time a trough from the other bobber arrives. This situation means that at a point like Q (where destructive interference is occurring), there is no motion of the water's surface. We have described how constructive and destructive interference occur at points P and Q as if these were isolated points. Instead of being isolated, however, the points P and Q lie, respectively, on lines of constructive and destructive interference. The diagram of Figure 13.6 is repeated in Figure 13.7, but lines have been added showing where constructive and destructive interference occur.

The particular pattern of lines for constructive and destructive interference shown in Figure 13.7 is valid only when the two bobbers are vibrating in phase with each other. We can see most easily why this statement is true by looking at the line of constructive interference that runs straight across the center of the diagram. Note that any point on this line is the same distance from bobber A as it is from bobber B. Since crests are created at each bobber simultaneously and since the crests travel at the same speed, they will arrive at any point on this line at the same time. Thus, there is always constructive interference at any point along this line. The two lines of constructive interference on either side of the central one occur because any point on those lines is exactly one wavelength farther from one bobber than from the other. The positions of the other lines of constructive interference are determined in the same way. That is, any point on one of the lines is an integral number of wavelengths farther from one bobber than from the other. Finally, we note that the distance from any point on a line of destructive interference (such as point Q) to one of the bobbers is exactly a half-integral number of wavelengths longer or shorter than the distance from the same point to the other bobber.

The number of lines of constructive and destructive interference that appear in an interference pattern (or, equivalently, how close together they are) depends on the ratio of the wave-

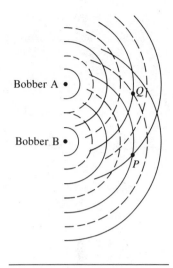

Figure 13.6. Wavefronts produced by bobbers A and B moving up and down at the water's surface in a ripple tank. The two bobbers are vibrating in phase with each other. Solid lines are wave crests, and dashed lines are wave troughs.

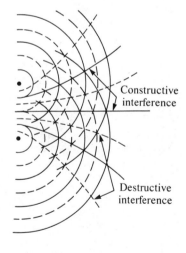

Figure 13.7. The interference pattern for two sources oscillating in phase with each other

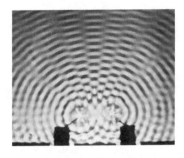

The interference pattern of waves generated by two sources oscillating in phase. The sources are about four wavelengths apart. (PSSC Physics, 2nd edition, 1965; D. C. Heath and Company with Education Development Center, Inc. Newton, MA)

length to the source separation. Note that the bobbers in Figures 13.6 and 13.7 are separated by a distance of about two wavelengths. As an illustration of this effect, Figure 13.8 shows the interference pattern for sources separated by only one wavelength in A and by four wavelengths in B.

As mentioned previously, the interference pattern shown in Figures 13.6 and 13.7 is valid only when the bobbers are vibrating in phase with each other. If the two bobbers are vibrating in antiphase with each other, then the lines of constructive and destructive interference will interchange.

This situation is shown in Figure 13.9, and we see that the line straight across the center of the pattern is now a line of destructive interference. Using the same line of reasoning we just used, we can see why this situation is true. Since the bobbers are oscillating in antiphase, a crest is produced by one bobber at the time a trough is produced by the other. This crest and trough will travel at the same speed over the same distance to reach a point on the center line at the same time and will cancel each other out. The rest of the argument follows in the same way, showing that when the difference in distances from some point to one bobber and from the same point to the other bobber is equal to an integral number of wavelengths, there will be destructive interference at the point.

In the previous discussion, we have seen that different interference patterns are produced for different phase relations between the two sources of waves. For each of the specific cases we considered, however, there was a definite phase relation between the sources, and this fact is important to recognize. If some definite phase relation between the sources is not maintained, then there will be no interference pattern. For every

Figure 13.8. The variation of the interference pattern with change in source separation to wavelength ratio

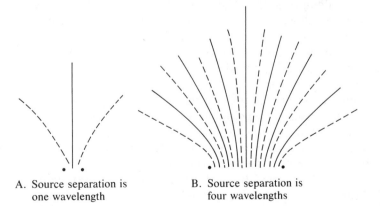

A. Source separation is one wavelength

B. Source separation is four wavelengths

phase relation between the sources, there will be a different pattern produced, but there will be a pattern! Because this idea of maintaining a definite phase between the two sources is so important to the production of an interference pattern, physicists have a special name for it: *coherence*. Thus, we say that two sources are coherent if they maintain a definite phase between them.

We have seen how interference is produced for water waves that are emitted from two coherent sources. Using the same ideas and techniques, we can understand the interference of sound waves. When a pure tone is broadcast through two speakers, we will have two coherent sources sending out waves. We easily can connect the speakers so that they vibrate either in phase or in antiphase with each other. (This phase relation can be changed by interchanging the wires on one of the two speakers.) The interference pattern set up in the room containing the two speakers can be detected easily. At points of constructive interference, there will be a loud sound, while at points of destructive interference, there will be no sound at all. We expect an interference pattern similar to the one produced with water waves, and we can look for this interference pattern by walking around the room. When we walk parallel to a line through the two speakers, we should find successive points of constructive interference about one wavelength apart. There is, however, a complication that might disturb this expected pattern. When the sound waves bounce off the walls of the room, they also can interfere with each other or with the waves coming directly from the speakers. Although this complication may disrupt the expected interference pattern, points of constructive and destructive interference in the room will be quite evident.

Obtaining coherent sources is the greatest problem we face when trying to observe interference with light waves. The basic reason for this problem is that we have no way of controlling the phase for a source of light waves. Therefore, it is impossible for us to have independent sources of light that are coherent. This difficulty does not mean, however, that we cannot obtain interference patterns with light; it just means that we have to use special methods to have two coherent sources.

The basic approach to obtaining two coherent sources of light is to use two parts of the same wave spreading out from a single source. Of course, these two parts of the wave have to move into the same region in order to interfere with each other. In addition, the two interfering waves have to appear as if they came from two sources that are close together. There are two

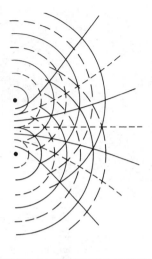

Figure 13.9. The interference pattern for two sources oscillating in antiphase with each other

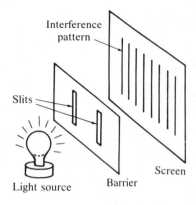

Interference pattern

Slits

Screen

Light source   Barrier

Figure 13.10. Slits in a barrier serve as two coherent sources of light for interference experiments.

ways of creating this situation. The first uses two slits in an opaque barrier placed in front of the light source. (See Figure 13.10.) When the light waves reach the barrier and pass through the narrow slits, they are diffracted, and they spread out. When the waves spread out, they form an interference pattern because the slits are serving as coherent sources.

The easiest way to observe this *two slit* interference pattern is with specially prepared slides. Usually, such slides have several different pairs of double slits with different distances between the slits. Thus, the different pairs of slits will give different ratios of source separation to wavelength. To use such a slide, we look through the pair of slits toward the source of light, much like we did with the single slit to observe a diffraction pattern. As we move the slide in front of our eyes and look through different pairs of slits, we can see the effect of changing the separation distance. For each case, there is an interference pattern superimposed on a diffraction pattern. The diffraction pattern occurs because the light is passing through narrow slits. The broad central diffraction beam (zeroth order beam) is broken into a series of dark and bright bands. This series of bands is the superimposed interference pattern. When the slits are close together (that is, when the slit separation is about the same as the width of each slit), then the central diffraction beam will be broken up into about three alternating dark and white bands. As the slit separation is increased, we can notice an increasing number of bright and dark bands in the central diffraction beam.

The second way we can observe light wave interference is with a thin film. When light is reflected off the front and rear surfaces of a thin film, we again get two coherent sources. In order to have the sources close together, we must have a thin film. The thin film, itself, can be made out of any type of transparent material. In addition to specifying what the film is made out of, we often specify the material on each side of it. The need for this specification becomes apparent when we recognize that one possibility is a thin film of air located between two pieces of glass. Other more common examples are a soap film in air and a film of oil or gasoline floating on the surface of water in a puddle.

In all of these cases of *thin film interference,* the thickness of the film determines the distance between the sources. Since the light waves reflected from the front and rear surfaces are the two sources, the thickness of the film also determines the phase between the sources. There is one additional factor, however. A light wave reflected from an interface where the speed of light increases has no phase change, but when the speed of light decreases across the interface, the light wave is inverted. (The light wave has a 180° phase change.) Note the similarity between this behavior of phase changes for reflected light waves and the behavior that occurred for reflected waves on strings. Recall from Chapter 8 that waves on a string were reflected from a point where the wave speed changed abruptly. One way we could have had an abrupt change in wave speed was to have had two different weight strings joined together. When the wave speed decreased across the junction, the reflected wave had a 180° phase change, and when the wave speed increased, there was no phase change for the reflected wave.

These are the fundamental facts that we will use to explain thin film interference. Now, we will discuss what we can observe for interference with a thin film, using as an example a thin soap film in air. For a film thickness that is exactly equal to three-fourths the wavelength of green light, the reflected waves from the front and back surface of the film will interfere constructively, giving a green color to the light. Light waves of other colors also will interfere, but this interference will not be as strongly constructive as it is for the green light. Thus, from white light incident upon such a film, the green light will be most strongly reflected.

We will use Figures 13.11 and 13.12 to describe this type of interference. In Figure 13.11, we see a greatly magnified cross-sectional view of the film along with the incident, refracted (transmitted), and reflected rays of the light. These rays have been drawn at slight angles so that they can be distinguished easily from one another. (The fact that the rays are not normal to the surface—they are at a slight angle—means the light travels a distance slightly greater than the thickness of the film in mov-

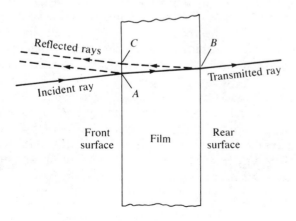

Figure 13.11. Incident, reflected, and transmitted rays of light for a thin film

Figure 13.12. Waves of light showing constructive interference for reflection from a thin film

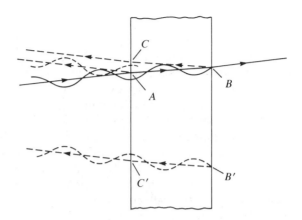

ing from one surface to the other. In this discussion, we will neglect that difference.) Thus, we will describe the distance from point A to point B as being equal to the thickness of the film. (The distance would meet this specification if the light were passing straight through the film.)

Figure 13.12 shows the same situation as Figure 13.11 with one addition and one modification. The addition is obvious: Transverse waves have been added to the rays. To understand the modification, we must proceed carefully. The ray and the superimposed transverse wave showing the light reflected from the second surface have been moved in order to prevent the confusion that would result from an overcrowded picture. (Of course, this movement may be a new source of confusion!) The thing you must recognize is that the wave that is pictured as moving from B' to C' and out actually moves from B to C and out.

Keeping in mind this displacement of the reflected beam, we will now use Figure 13.12 to study the details of thin film interference. First, we note that the wavelength of the light is shorter inside the film. This decrease in wavelength occurs because light moves more slowly inside a soap film than it does in air. Next, we note the phase of each reflected wave. At the front surface where the speed of the transmitted wave decreases, the reflected wave is inverted. Since the transmitted wave on the far side of the rear surface moves faster than the wave incident on that surface, the wave reflected there is not inverted. As these two reflected waves leave the front surface (moving to the left in Figure 13.12), they will interfere. When they are in phase with each other, there is constructive interference, and when they are out of phase, there is destructive interference. The thickness of the film and the wavelength of the light determine whether one of these situations occurs. Figure 13.12 has been drawn for the case when constructive interference is occurring. Both of the reflected waves leaving the front surface are in phase with each other; at the instant of time pictured, they both have crests at the surface. The wave that has reflected off the back surface has traveled a distance equal to twice the thickness of the film farther than the other wave. There is also a 180° phase shift between the waves because of the differences between speed changes across the front and rear surfaces. We recall that a 180° phase shift is the same as a shift of one-half a wavelength.

Putting these two facts together shows us that one wave has fallen behind the other one by an effective distance of twice the thickness of the film minus one-half the wavelength. If this effective distance is equal to an integral number of wavelengths, then the two reflected waves leaving the front surface will be in phase with each other, and we will have constructive interference. On the other hand, if this effective distance is equal to a half integral number of wavelengths, then the two emerging waves will be out of phase, and we will have destructive interference. Using variables, we can write these ideas concisely as

$$\text{effective distance} = 2t - \frac{1}{2}\lambda \tag{13.1}$$

$$\text{effective distance} = n\lambda \quad \text{(constructive interference)} \tag{13.2}$$

$$\text{effective distance} = (n - \frac{1}{2})\lambda \text{ (destructive interference)} \tag{13.3}$$

In these equations, t = film thickness, and n = zero or a positive integer. We can solve these two relations for the film thickness required to have constructive and destructive interference respectively.

$$t = \frac{1}{2}(n + \frac{1}{2})\lambda \qquad \text{(constructive interference)} \tag{13.4}$$

$$t = \frac{1}{2}n\lambda \qquad \text{(destructive interference)} \tag{13.5}$$

In these equations, n = zero or a positive integer.

We can make two predictions based on these relations.

1. To change from constructive to destructive interference or vice versa, the film thickness must change by one-fourth a wavelength.

2. When the thickness of the film is zero, $n = 0$, so we always will have destructive interference. This destructive interference happens for all wavelengths.

Based on these ideas, we can expect to see the following things happen with a thin film (for example, a soap film) in air. When the film is very thin (the thickness is approximately zero), there will be no light reflected from the film, and the film will appear to be black. The reason for this black appearance, of course, is because destructive interference is occurring for all wavelengths of light. Now, suppose we could gradually increase the thickness of this film. The first color we would see would be blue, because it has the shortest wavelength. This blue color would appear when the film had the proper thickness to give constructive interference for blue light. According to Equation 13.4, this situation occurs for a film thickness of 100 nm, which is one-fourth the wavelength of blue light. As the film thickness continues to increase, the color of the reflected light will change.

We also find from Equation 13.4 that the first thickness to give constructive interference for red light is 175 nm. In both these cases, of course, other wavelengths of light have partial constructive interference. Thus, the colors that we see will not be pure; in the words of Chapter 12, the colors are unsaturated. The effect of dispersion, however, helps separate the various colors in the constructive interference pattern. This reinforcement happens because the different wavelengths of light are separated more widely in the film than they are in air. (Recall that the wavelengths of both red and blue light are shortened in the soap film, but since the blue light slows down more than the red light, the wavelength of the blue light will be shortened more than the red light's wavelength.)

As the film gets progressively thicker, we will observe the color change from blue through the spectrum to red and then back to blue again. A continued increase in thickness will change the color from blue through the spectrum to red again. The first time the color changes from blue to red, the value of n in Equations 13.1 through 13.5 is zero. For the second pass through the spectrum, $n = 1$, and the thickness of the film is equal to three-fourths the wavelength of each color. Note that Equation 13.4 predicts an infinite number of film thicknesses for constructive interference (as n takes on all possible values). This prediction suggests that when we shine white light on a film, we should get only colors reflected from any thickness of that film.

However, as the film gets thicker, it will provide for constructive interference of one color with one value of n and constructive interference for another color with a different value of n. The addition of these colors and many others will give white light. Thus, we see colors in light reflected from films only when the films are thin so that constructive interference is occurring for only the lowest few values of n.

We have seen how two sources of coherent waves produce an interference pattern. Increasing the number of sources (but keeping them equally spaced) enhances the constructive interference. As we add more sources, each region where constructive interference occurs gets narrower and stronger because more waves are superposing. (As mentioned earlier, slits in an opaque barrier serve the same function for light as do individual sources for other waves.)

A piece of glass having a large number of equally spaced "lines" (which are transparent slits) separated by opaque bands is called a *grating*. A typical grating has about 5,000 lines per centimeter, so it is the ultimate in multiple slit devices for light. It often is called a diffraction grating, and the light does indeed diffract as it passes through each slit. But interference also occurs, so we simply will call the object a grating. By using a grating, we can separate the colors of the spectrum as well or better than by using a prism. However, just as for diffraction, red light is bent more than blue. (This similarity in color separation is probably a major reason for using the name diffraction grating.) We can use a grating to observe various light sources. For an incandescent filament, we see the complete spectrum, and the various colors produced by a gas discharge tube are separated nicely.

Scattering

A wave is scattered when it is reflected off many small objects. These objects can be pilings (posts) for water waves; buildings or pillars for sound waves; or very small particles for light waves. The difference between reflection and scattering is that a reflected wave still maintains the same type of coherence that the incident wave had. Coherence is lost for scattered waves because they are reflected off many small points located at different positions in space. Consequently, the wave reflected from each individual object is a circular or spherical wave (depending upon whether the wave is a water wave, or a light or sound wave). In order to get such circular or spherical wavefronts, the objects doing the scattering must be small. We find, however, that there is a fundamental difference between two cases. The first occurs when the objects doing the scattering are smaller than the wavelengths being scattered, and the second happens when the objects are larger than the wavelengths. Thus, we divide the subject of scattering into two categories: small object scattering and large object scattering. Even for large object scattering, however, the sizes of the objects will not be much greater than the wavelengths of the waves being scattered.

The reason for separating scattering into these two categories becomes evident when we look at what happens to waves of many different wavelengths that are being scattered at the same time. This is the situation, of course, when we consider the scattering of white light. When we have small object scattering for such a mixture, we find that shorter wavelength waves are scattered more effectively than longer wavelength ones. On the other hand, when the objects are bigger than any of the wavelengths involved, all the waves are scattered equally well. The marching soldier metaphor that we used in our description of refraction and dispersion of waves will also work here. That is, we can represent a mixture of waves of different wavelengths by a file of soldiers of different sizes. The shorter soldiers will act like the shorter wavelength waves, and vice versa. Now, imagine this squad marching across a field that is strewn with boulders. For the tall soldiers, the boulders are about knee-high, but the boulders are waist-high for the shorter soldiers. To complete the scattering analogy, we also must imagine that when a soldier encounters a boulder he cannot step over, he must turn to one side. Since the taller soldiers can step over many of the boulders they meet, they do not have to turn aside (that is, be scattered) nearly as often as the shorter soldiers do. On the other hand, if the boulders are all bigger than the tall soldiers, then both sizes of soldiers will be scattered equally.

We look at light as the prime example of both types of scattering. Both types can be observed with water waves, however. The only difficulty is that unlike light, we usually do not see a mixture of wavelengths at the same time with water waves. But we can observe the amount of scattering that occurs as the wavelengths of the water waves change. One way to observe this scattering is to find a nice spot on the shore near a set of scattering objects. These objects might be some pilings in the ocean or some reeds along the edge of a small pond. Then, on a day when the water is calm but the wind is starting to blow, we can go out to this position and make our observations. As we saw in Chapter 7, when the wind starts to blow, the waves will build in height. But the height is not what is important. We are interested in the wavelength, which, of course, also increases. As we start our observations (when the waves are small), we will see significant scattering. As the waves build in length, less and less scattering will occur. Finally, when the waves are much longer than the size of the scattering objects, the waves will pass right by without even noticing the scattering objects.

As we have mentioned, the scattering of light waves is an important example of this wave behavior. When the light is scattered off objects bigger than the wavelength of any visible light, all wavelengths will be scattered equally. This is the case for most dust, smoke, or water particles suspended in the air. A typical size for water particles is compared to the wavelengths of red, green, and blue light in Figure 13.13. Thus, when white light is scattered by water particles—or by dust or smoke particles—all colors are scattered equally well, and the scattered light is white.

The other scattering case also occurs for light. Both the molecules of air and very fine volcanic dust are scattering objects that are smaller than the wavelengths of visible light. A comparison of the sizes of air molecules to wavelengths of visible light is made in Figure 13.14. When white light passes through the air, the blue end of the spectrum is scattered better than the red end. This effect is enhanced when the air contains fine volcanic dust in suspension. In either case, the scattered light is

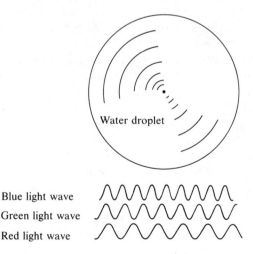

Blue light wave
Green light wave
Red light wave

Figure 13.13. A size comparison between the wavelengths of visible light and a typical water droplet in a stratus cloud

Figure 13.14. A size comparison between the wavelengths of visible light and air molecules

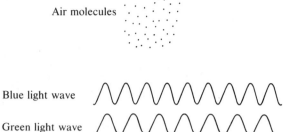

Air molecules

Blue light wave

Green light wave

Red light wave

blue, and the original light beam turns red. In the next chapter, we will discuss some striking atmospheric effects that are caused by this preferential scattering of short wavelength blue light.

Polarization

We have seen how a transverse wave moving along a string causes the string to move first to one side and then to the other. In our previous discussions, the direction the string moved as the wave passed by was determined by how we moved the end of the string when we started the wave. For example, if we moved the end of the string up and down, then the wave produced an up and down motion of the string. In a similar way, the motion of the string could be restricted to the horizontal direction by moving the string initially back and forth horizontally. When the string is restricted to moving in only one direction such as either up and down or back and forth horizontally, then we say the wave is *polarized*. (Notice that when the wave is polarized, the movement of the string is restricted to lie in a certain plane, so we sometimes refer to this condition as *plane polarization*.) When we move the string up and down, the wave traveling along the string causes the string to move in a vertical plane, and we say the wave is polarized with its plane of polarization vertical. If the string only moves horizontally, then the wave is polarized in the horizontal plane. We, of course, could restrict the motion to any other plane with an orientation other than vertical or horizontal, and in such a case, the wave also would be polarized.

Since I want to be able to describe the polarization of a wave by means of diagrams, let me illustrate the type of diagram I will use. The information we wish to convey with such a diagram is the vibration direction of each point of the string as the wave

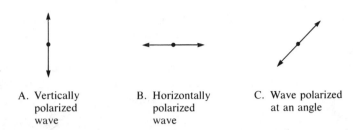

A. Vertically
   polarized
   wave

B. Horizontally
   polarized
   wave

C. Wave polarized
   at an angle

Figure 13.15. Diagrams showing direction of polarization for transverse waves

passes, that is, the direction of polarization. Conveying this information is very difficult to do with a drawing that looks at the string from the side; we can convey this information much more easily if we draw a picture that looks along the string. So, we draw a picture where the string is coming straight out of the page. In such a drawing, of course, the string will be only a dot on the page. But we want to show the direction of vibration for the string, and we easily can show this direction by adding a line with arrows on each end. For a wave on the string polarized in a vertical plane, the arrows will point up and down. For other directions of polarization, the arrows will be changed accordingly. In Figure 13.15, pictures A–C show three different polarized waves on a string using this type of diagram.

From the above discussion, it should be clear that only transverse waves can be polarized. The motion of the medium during the passage of either a longitudinal wave or a water wave cannot be restricted in the same way it can be for a transverse wave. But not all transverse waves are polarized. I also will use the string to describe an unpolarized wave. As an unpolarized wave moves along the string, the motion of the string is not restricted to one direction. It may be a little difficult to visualize this situation if we have in mind only a few crests and troughs in such a wave. If we think of a wave that has hundreds or thousands of undulations in it, however, then it becomes easier to imagine some of these undulations moving the string up and down and some of them moving it horizontally. Indeed, for our very long unpolarized wave, the string moves in every direction as it passes by. A *polarization diagram* for this unpolarized wave is shown in Figure 13.16. Strictly speaking, of course, this diagram should have arrows pointing in every direction. The use of four different angles for these double-arrowed lines shows clearly that the wave is not polarized; therefore, it is *unpolarized*.

In the previous discussion, we had either polarized or unpolarized waves, depending upon the way the end of the string was moved. Now, I want to describe how we can change an unpolar-

Figure 13.16. The polarization diagram for an unpolarized wave

ized wave on a string into a polarized one. We again must try to visualize a very long wave moving down a string. For this case, however, the string passes through a slot in a barrier. A sketch of this situation is shown in Figure 13.17, where the barrier has a vertical slot that the string passes through. As the unpolarized wave reaches this barrier, any vibrations in the vertical direction will pass through the slot. However, any vibrations in the horizontal direction will be stopped; the barrier absorbs the energy of these waves. How about vibrations at an angle? A vibration at an angle means, for example, that the waves move the string both up and to the right at the same time and then down and to the left during the next half-cycle of the wave. That part of the movement which is in the vertical direction will pass through the slot, while that part of the movement in the horizontal direction will be absorbed. Thus, on the incident side of this barrier, we have an unpolarized wave on the string. (I have drawn this side with a larger diameter because the string vibrates in every direction as the wave moves along.) The wave transmitted beyond the barrier, however, is polarized in a direction parallel to the slot. This mechanism for polarizing the wave is termed *selective absorption,* and the device is called a *polarizer.* This discussion concerning polarization of waves on a string is interesting because we can actually see—or at least visualize in our minds—these waves as they move along the string.

We have seen how a slot in a barrier can be used to produce polarized waves from unpolarized ones by means of selective absorption. We can also use the same device (that is, a slot in a barrier) to test a wave to see if the wave is polarized. (This seems to be an unnecessary exercise because we actually can observe whether waves on a string are polarized or not. Shortly, however, we will use these same techniques with light waves, which we cannot observe directly.) In order to use a slotted barrier as a polarization *analyzer,* we also must have the string passing through the slot. Now, however, we rotate the barrier, as shown in Figure 13.18, and we observe the wave on the far side of the

Figure 13.17. An unpolarized transverse wave becoming polarized after passing through a slot in a barrier

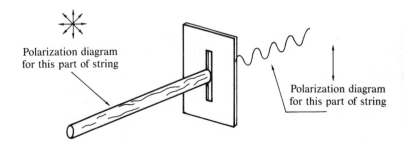

Polarization diagram for this part of string

Polarization diagram for this part of string

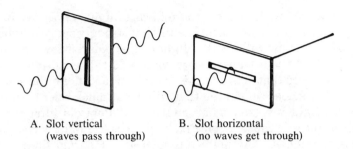

Figure 13.18. Rotation of a slotted barrier to detect polarization.

A. Slot vertical
(waves pass through)

B. Slot horizontal
(no waves get through)

barrier. If the incident wave is unpolarized, then there always will be a wave passing through the slot as the barrier is rotated, and this transmitted wave will have a constant amplitude. This situation does not happen if the incident wave is polarized, however. In that case, the size of the wave that passes through the slot will depend upon the relative orientations of the slot and the plane or direction of polarization of the original wave. When the two are parallel, the wave passing through the slot will have the same amplitude as the incident wave. When the barrier is rotated by 90° from this position, however, there will be no wave passing through the slot. (The wave will all be absorbed). Thus, if the amplitude of the transmitted wave changes as we rotate the barrier, then the incident wave must have been polarized. Finally, we note that if the string passes through slots in two successive barriers (see Figure 13.19), then the first slotted barrier will act as a polarizer, and the second slotted barrier can act as an analyzer.

Figure 13.19. The use of two slotted barriers, one as a polarizer and the other as an analyzer

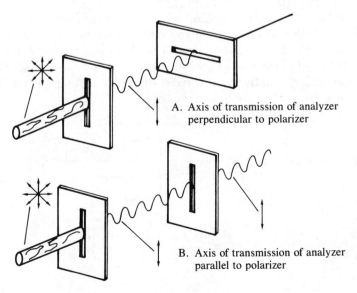

A. Axis of transmission of analyzer
perpendicular to polarizer

B. Axis of transmission of analyzer
parallel to polarizer

Now, we wish to apply these ideas to light waves. We have discussed a number of experiments that strongly suggest light is a wave. We now will see that light can be polarized, which tells us that light is a transverse wave. To do this experiment, we need a device that acts like the slotted barrier did for the string. That is, we need something that will act as a polarizer for unpolarized light waves or as an analyzer. A Polaroid filter serves this purpose because it selectively absorbs the light waves that pass through it. Any material (such as a Polaroid filter) that selectively absorbs light waves is called a *dichroic* material. For light waves, a piece of Polaroid material acts exactly as the slotted barrier did for waves on a string. That is, if we shine a light through a piece of Polaroid material, rotate the Polaroid material, and see that the transmitted light changes its intensity, then we know the incident light is a polarized beam. Most light sources, however, are not polarized.

But we easily can see that light can be polarized by passing a beam successively through two pieces of Polaroid material. By holding either one of the pieces fixed and by rotating the other one, we will see the light beam emerging from the second piece of Polaroid material change from bright to dark and then back to bright again during every half revolution of the piece of Polaroid material we are turning. This experiment is easy for us to understand because exactly the same thing happened with the slotted barriers and the waves on the string. That is, the first piece of Polaroid material serves as a polarizer, so the light beam between the two pieces of Polaroid material is polarized. When the second piece of Polaroid material is rotated, every quarter of a revolution, its axis of transmission will change from being parallel with the polarized light to being perpendicular to it. The light wave can pass through the second piece of Polaroid material when its direction of polarization is parallel to the axis of transmission, and the light wave cannot pass through when the two are perpendicular to one another.

Figure 13.20. The use of two pieces of Polaroid material for a polarizer and analyzer (Axes of transmission are parallel.)

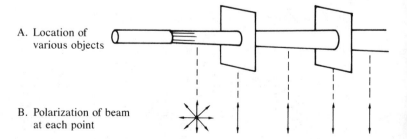

A. Location of various objects

B. Polarization of beam at each point

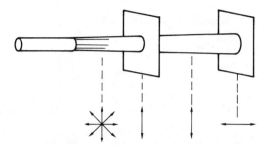

Figure 13.21. The use of two pieces of Polaroid material for a polarizer and analyzer (Axes of transmission are perpendicular.)

Figure 13.20 shows a diagram of this experiment. In picture A, we see a side view of the experiment showing the location of the source of light and the two pieces of Polaroid material. The polarization diagrams and the directions of the transmission axes of each piece of Polaroid material are shown in picture B. In this particular figure, the axis of transmission of the first piece of Polaroid material is vertical, so the polarization direction of the light beam between the two pieces of Polaroid material is also vertical. Since the second piece of Polaroid material also has a vertical axis of transmission, the light will pass through this piece of Polaroid material. Figure 13.21 shows the same experiment, except the second piece of Polaroid material has been rotated by 90° so that no light passes through it; its axis of transmission is perpendicular to the polarization direction of the light wave incident upon it. As stated earlier, this experiment shows us that light is a transverse wave.

Using dichroic material to polarize a light beam is very convenient. This method, however, is not the only way to obtain polarized light. Light that is reflected at an angle off a nonmetallic surface also will be polarized. This light, however, may be only *partially polarized*. By this statement, we mean that the light wave vibrates in every direction, but its amplitude in one direction is larger than that in the perpendicular direction.

Figure 13.22 shows a polarization diagram for a partially polarized beam of light. If we pass this beam of light through a Polaroid filter, rotate the filter, and observe the transmitted beam, then we will see the intensity of the beam change from bright to dim (but not totally dark). This reaction is usually what happens when we look at light reflected off a nonmetallic surface through a piece of Polaroid material and then rotate the piece of Polaroid material. The intensity of the light changes, showing that the reflected light is partially polarized. The direction of the light beam's larger amplitude is parallel to the surface off which the light has reflected, as shown in Figure 13.23.

Figure 13.22. The polarization diagram for a partially polarized light beam

Figure 13.23. Polarizing light by reflection off a nonmetallic surface. (Total polarization occurs only at Brewster's angle.)

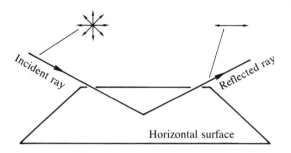

Incident ray

Reflected ray

Horizontal surface

At one particular incident angle (and, therefore, reflected angle) the light will be totally polarized. This angle (called *Brewster's angle* after David Brewster, who first discussed it in 1815) varies from material to material but usually has a value of about 50°. Polarized sunglasses work so well in eliminating glare because they absorb much of the light reflected from horizontal surfaces. The axis of transmission of the Polaroid material in the sunglasses is always vertical. This selective absorption of reflected light can be very striking when one is looking down into a pond of water. Eliminating the reflected glare allows one to see into the water much more clearly. Light reflected from a metallic surface, however, is not polarized. Thus, a mirror (which has a metal backing) will not show polarization of the light reflected from it.

The process of scattering is another way in which light can become polarized. This process is also a result of the fact that light is a transverse wave. As with reflection, maximum polarization occurs at one angle. Figure 13.24 shows why this angle is 90°. In this diagram, the light beam coming from the source through the scattering material is unpolarized. Therefore, of course, the beam has vibrations in every direction perpendicular to the direction in which it is moving. Some of these vibrations will be in the direction in which the scattered beam moves and, therefore, would produce longitudinal waves in that beam. Light is not a longitudinal wave, so these vibrations cannot create the scattered wave. Those vibrations in the incident wave that are perpendicular to the ones just discussed also will be transverse for the scattered wave. Thus, only these vibrations can produce the scattered wave, and we see that the scattered wave is polarized.

There are two other interesting effects involving polarized light that I would like to describe. The first occurs with certain crystals, which are called *doubly refracting* or *birefringent* crystals. A light beam entering such a crystal splits into two beams.

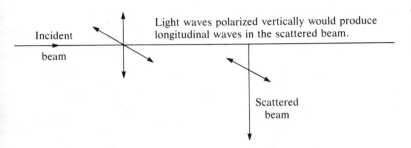

Light waves polarized vertically would produce
longitudinal waves in the scattered beam.

Incident
beam

Scattered
beam

Figure 13.24. Polarizing light
by scattering

Both beams are polarized; their directions of polarization are
perpendicular to each other. This situation is shown in Figure
13.25. As indicated, one of the beams passes straight through the
crystal. The other is deviated from this path and emerges to one
side of the first beam. Thus, if we rotate the crystal and look at
the emerging beams, one of them will appear to stand still, and
the other will rotate around the first one. For obvious reasons,
we call the first beam an *ordinary ray* and the second one an
extraordinary ray. Calcite is a crystal that is often used to
observe this behavior because the effect is very large for this
mineral.

The second interesting effect can be observed using water in
which some sugar has been dissolved. When polarized light is
passed through such a solution, the direction of polarization is
rotated. The amount of rotation depends upon the length of the
path of light in the water and also upon the wavelength of the
light. That is, different wavelengths of light are rotated by different amounts. Thus, if we pass polarized white light through a
sugar solution and observe the transmitted light with a piece of
Polaroid material, we may see different colors. Further, as this
analyzing piece of Polaroid material is rotated, the colors will
change. This rotation of polarization direction is shown in Fig-

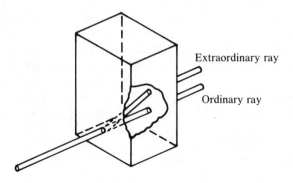

Extraordinary ray

Ordinary ray

Figure 13.25. A beam of
unpolarized light passing
through a birefringent crystal

Figure 13.26. Rotation of polarization direction by an optically active material

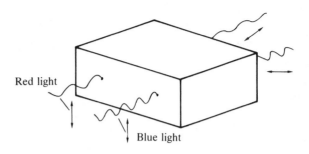

Red light

Blue light

ure 13.26, where the differing amounts of rotation for red light and blue light have been indicated. Materials that rotate the plane of polarization, such as the sugar solution, are called *optically active*. Certain types of plastics and some crystals are also optically active. (Plate 7 in the color section of this text shows an example of polarized light passing through a piece of optically active stressed plastic.)

We have discussed a number of different types of materials that affect the polarization of light. All of these effects are produced by atoms acting on the light, so we cannot directly observe the processes occurring. However, using the information we gained from the slotted barriers acting on the waves moving along the string, we can understand how such processes can happen.

Summary

Diffraction, interference, scattering, and polarization are four new types of wave behavior discussed in this chapter. In each case, we first examined the behavior using waves we actually can see, such as water waves or waves on a string. Then, we saw how light and sound waves act.

Diffraction is the bending of waves around a barrier. The easiest way to observe this effect is to use two barriers to form an opening that has a width about equal to the wavelength of the waves passing through it. In such an experiment, we see the breakup of the diffracted wave into several distinct beams, and the extent of the diffractive spreading depends on the ratio of the wavelength to the opening size. We also can observe diffraction when a wave passes by a barrier, but the effects are not as obvious. Diffraction occurs for all types of waves; we discussed examples of it for water, light, and sound waves.

Interference occurs when waves from two (or more) coherent sources move into the same region. The interference pattern depends on both the distance between the sources compared to the wavelength and the relative phase of the sources. There are two easy ways to observe interference with light waves. In the first, light from a single source is passed through two parallel slits. These slits then serve as coherent light sources, and the interference pattern is seen as parallel bands of light and dark. The second way of observing interference effects with light occurs when light is reflected from a thin film. In this case, the front and rear surfaces of the film serve as coherent sources, and the thickness of the film determines which wavelengths of light constructively or destructively interfere.

Scattering is the term used to describe the incoherent reflection of waves off many small objects. This phenomenon can happen with all types of waves, but we see the effects most often with light waves. We recognize two types of scattering, depending on the size of the scatterers compared to the wavelength. For scattering off objects larger than the wavelengths being scattered, all waves are scattered about equally. However, wavelength-dependent scattering occurs when waves are scattered off objects smaller than the wavelengths.

The concept of polarization has meaning only for transverse waves. A transverse wave is polarized when the vibrations of the wave are limited to a single direction. The fact that light can be polarized and sound cannot shows that they are, respectively, transverse and longitudinal waves. A wave on a string can be polarized by having the string pass through a slot in a barrier. Such a slotted barrier also can be used as an analyzer to check whether a wave on a string is polarized. Dichroic materials, such as Polaroid filters, perform the same functions for light waves. Light waves also are polarized when reflected from a nonmetallic surface, although these waves are only partially polarized for all except one incident angle.

Suggested Readings

The topics of diffraction and interference are explored for both sound and light in *Sound Waves and Light Waves,* by Winston E. Kock (Garden City, N. Y.: Anchor Books/Doubleday, 1965). He has some interesting techniques for being able to "see" diffraction and interference of sound waves, and he shows some of the resulting pictures.

Thin film interference of light reflected from a soap film is discussed in some detail in "The Amateur Scientist" column, *Scientific American,* September 1978. Columnist Jearl Walker discusses the creation of thin film interference in much the same way that we have done here. In addition, he has several nice photographs of soap films exhibiting thin film interference, and he describes how to mix a solution that will yield strong soap films.

Review Questions

1. Which of the following effects make it impossible to take completely sharp pictures with a pinhole camera?

   a. Diffraction

   b. Interference

   c. Polarization

   d. Dispersion

   e. Refraction

2. We often talk about light in terms of rays but rarely mention rays in connection with sound. Which difference between light and sound allows us to think of light waves moving in straight lines more easily than sound waves?

   a. Light is a transverse wave, but sound is a longitudinal wave.

   b. The speed of light is much greater than the speed of sound.

   c. The frequency of visible light waves is much higher than the frequency of audible sound waves.

   d. The wavelength of visible light waves is much shorter than the wavelength of audible sound waves.

   e. We see light, but we hear sound.

3. White light passes through a single narrow slit and falls on a white screen. We observe which of the following on the screen:

   a. a broad, central, white band with narrower white bands on either side.

   b. a broad, central, white band with narrower colored bands on either side. On one side, the bands are blue, and on the other side, the bands are red.

   c. a number of narrow, white bands.

d. a number of colored bands that alternate between blue and red.

e. a number of colored bands with all colors of the spectrum represented in order.

4. A tsunami with a wavelength of 250 kilometers is heading due north toward the islands shown in Figure 13.27. Each island is about 200 kilometers long. To minimize panic, but to provide maximum safety for the inhabitants, the authorities should

a. warn only the people on island A because the interference between waves scattered off islands B and C will cancel.

b. warn only the people on island C because the interference between waves scattered off islands A and B will cancel.

c. warn only the people on islands A and B because island C will be shielded by island B.

d. warn the people on all three islands because the tsunami can diffract around island B and strike island C as well as islands A and B.

e. not warn anyone because the islands are so small compared to the wavelength of the tsunami that the tsunami will pass right by.

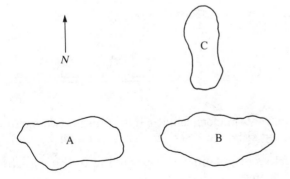

Figure 13.27. For use with review question 4

5. Which of the following is not capable of producing an interference pattern?

a. Two loud speakers that are connected to the same signal generator

b. Two light bulbs behind separate slits that are connected to the same power source

c. Two small slits illuminated with a laser

d. Light reflected from both surfaces of a soap film

e. Light reflected from the two touching surfaces of different microscope slides that are on top of one another

6. Figure 13.28 shows waves in a ripple tank that are passing through two slits in a barrier. The solid lines are crests, and the dashed lines are troughs. Therefore, at points A, B, and C, there will be

a. constructive interference, and the water will be still.

b. constructive interference, and the water will be choppy.

c. destructive interference, and the water will be still.

d. destructive interference, and the water will be choppy.

e. alternating constructive and destructive interference, so the water will be choppy.

Figure 13.28. For use with review question 6

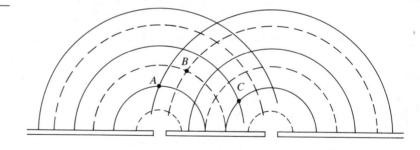

7. Coherent, monochromatic light falls on two narrow, closely spaced slits. On a screen that is a meter beyond,

a. two bright lines corresponding to the two slits will be seen.

b. two broad bands from the two slits will be seen.

c. nothing will be seen because light cannot get through narrow slits.

d. a series of bright lines with dark lines between them will be seen.

e. a complete spectrum created from the interference of the light from each slit will be seen.

8. When the dimensions of scattering particles are smaller than the wavelength of blue light,

   a. red light is scattered more effectively than blue light.

   b. both red and blue light are scattered about equally and better than green light.

   c. both red and blue light are scattered about equally but not as well as green light.

   d. blue light is scattered more effectively than red light.

9. We see a white cumulus cloud because light is scattered from the water droplets in the cloud. Since the clouds are white, we can infer

   a. that water droplets have a white color.

   b. that there are equal numbers of red, green, and blue water drops in the cloud.

   c. that the water droplets are smaller than the wavelengths of visible light.

   d. that the water droplets are larger than the wavelengths of visible light.

   e. nothing. (Not enough information is given.)

10. Polarized light may *not* be produced by

    a. selective absorption.

    b. double refraction.

    c. scattering.

    d. total internal reflection.

    e. reflection from a glass surface.

11. When laser light shines through a Polaroid filter, which of the following properties changes?

    a. Frequency

    b. Speed

    c. Coherence

    d. Wavelength

    e. None of the above

14

Atmospheric Phenomena

Important Concepts

- Rainbows are a result of dispersion, which occurs when light is refracted as it enters and leaves each individual drop in the rain shower producing the bow. The light is reflected once inside the drop for the primary bow and twice for the secondary bow.

- Wavelength-dependent scattering of sunlight off the gas molecules in the atmosphere produces the blue sky, red sunsets, and purple mountains.

- Wavelength-independent scattering of sunlight off water droplets larger than the wavelengths of visible light give us white clouds and fog.

- A corona is a thick circle of light around a source that is caused by the diffraction of light around water drops in a fog or mist. The outer and inner edges of the corona are colored red and blue, respectively, because longer waves are diffracted through larger angles.

- Refraction of light passing through rod-shaped ice crystals produces halos. A halo is a thin circle of light around a source with a bigger diameter than a corona. Since blue light is bent more than red light by refraction, the outer and inner edges of a halo are blue and red, respectively.

- Mirages are created by the refraction of light moving through regions of air having different temperatures.

299

■ Sun dogs, sun pillars, and sub suns are rarer atmospheric phenomena that occur when ice crystals are suspended in the air. Reflection of sun light is the mechanism for producing a sun pillar (a pillar of light above the sun) and a sub sun (an image of the sun below the horizon). Refraction produces a sun dog (an image of the sun to one side of the sun) in much the same way as a halo.

We can observe a number of interesting effects in the atmosphere that can be explained using light waves. All of these effects serve as examples of some type of wave behavior that we have discussed. Thus, the fact that these atmospheric events occur reinforces our belief that light is a wave. Some of the things we talk about will be familiar to you; everyone has seen a rainbow or a red sunset. Other events are less common, but just as striking. Having learned about them here, you may be more likely to see some of these rare happenings yourself because you will know when and where to look for them.

The Rainbow

You probably have heard of the old legend that claims there is a pot of gold at the end of the rainbow. Since legends like this one have a certain charm, we always feel a bit of sadness when they are proven wrong. We will see that—in spite of its apparent reality—the rainbow does not exist at some location in space. Rather, the rainbow is an image of the sun. The full spectrum of colors appears in this image for the same reason that light passing through a prism is broken into a spectrum. That is, there is refraction by a dispersive medium.

A rainbow is produced when light is refracted and reflected by small drops of water in the atmosphere. Light entering one side of each raindrop is refracted and dispersed. The light then strikes the back of the raindrop, where the light is reflected to the other side. As the light leaves this other side, it is refracted and dispersed again. Figure 14.1 shows the light being separated into the full spectrum as it moves through a single raindrop. Since the two refractions and the reflection always occur at an interface between air and water, the total angular change of direction of each wavelength of light is always the same. Further, the dispersion produces a different angle for each color of the spectrum.

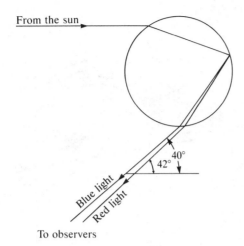

From the sun

40°
42°

Blue light
Red light

To observers

Figure 14.1. The paths of light rays through a raindrop to form a rainbow

In order to observe this light, we stand with our backs toward the sun so that we are facing the *antisolar point* (a point in space directly opposite the sun). We then look at the appropriate angles away from this direction to view the entire spectrum.

The extremes of the angles for viewing the rainbow are 42° for red and 40° for blue. (Note that our previous observation about blue light being bent more than red is valid here also. The amount of bending or change in direction is 180° minus this angle, so red is bent through 138°, and blue is bent through 140°.) As we can see from Figure 14.1, we would have to stand in different places in order to see the blue light or the red light from this single raindrop. When we observe a rainbow in the sky, we see the various colors coming from different raindrops. For each raindrop, however, the angles through which the light is bent for each color are the same as those shown in Figure 14.1. Since these angles are the same for raindrops high in the sky as well as for raindrops to the right and left, we get the familiar bow shape for the refracted and reflected light.

Figure 14.2 shows how this bow shape appears for a person looking at a raindrop-filled sky. For each person, the rainbow will lie between the angles of 40° to 42° away from the antisolar point. All the directions falling within these angles form a cone with the observer at the top (or vertex) of the cone. Thus, whenever the person looks anywhere along the side of this cone (and there are raindrops located at that position), he will see the colors of the spectrum. The quality of the rainbow (that is, how vivid the colors appear and how distinct the rainbow looks)

Figure 14.2. Viewing rainbows in the sky

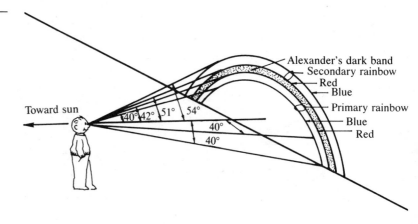

depends upon the size of the raindrops that are refracting and reflecting the light.

Sometimes, when conditions are just right, a second rainbow can be seen outside of the first one. This bow, called the *secondary rainbow,* has the order of its colors reversed from those in the *primary rainbow.* This secondary rainbow is formed by light being refracted and reflected from raindrops in the same shower producing the primary bow. However, in the secondary bow, the light makes two reflections inside each raindrop, rather than just the one reflection that occurs in the primary bow. Figure 14.3 shows the path of light within each raindrop for producing the secondary bow. From this figure, we can see that the second reflection of the light inside the raindrop causes the reversal of the colors in the dispersed spectrum compared to the

Figure 14.3. The paths of light rays through a raindrop that create the secondary rainbow

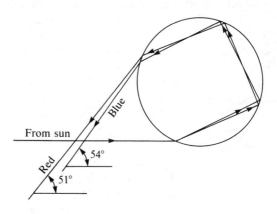

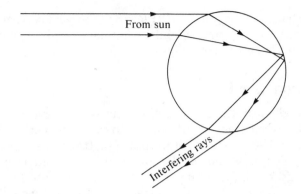

From sun

Interfering rays

Figure 14.4. Light rays that interfere to produce supernumerary bows

primary bow. It is theoretically possible to have higher order rainbows that would be created when the light is reflected even more times inside each raindrop. While these higher order rainbows have been observed in the laboratory using artificially produced drops, such higher order bows rarely are seen in the sky.

There are other observations we can make about the sky when we see a double rainbow. First, the region between the two bows will be darker than the rest of the sky. This region is named *Alexander's dark band* after a Greek philosopher, Alexander of Aphrodisias, who lived in the time around 200 A.D. and who first described the effect. This dark band occurs because raindrops do not reflect or refract light into this direction. Next, we note that the sky on the side of each bow away from Alexander's dark band is lighter. These "lighted sides" of each bow occur because light can be reflected and refracted into these directions. Finally, if we look carefully when conditions are just right, we can see extra, very thin, colored arcs below the primary bow. These arcs are called *supernumerary bows* (or *supernumerary arcs*) and arise because of interference rather than dispersion. The interference occurs between pairs of light rays that move through the raindrops in the same way that the rays forming the primary and secondary bows do. However, as shown in Figure 14.4, two rays of light can enter a raindrop at different points and leave in the same direction. The paths of these two rays through a drop have different lengths. Therefore, light waves that enter the drop in phase will have a phase difference when they leave the drop. This phase difference will depend upon both drop size and wavelength. For just the right conditions, constructive interference will occur, and colored, supernu-

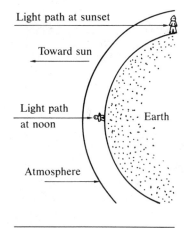

Light path at sunset

Toward sun

Light path at noon

Earth

Atmosphere

Figure 14.5. The difference in atmospheric thickness at noon and at sunset (or sunrise)

merary bows will be produced. (Plate 8 in the color section of this text shows both a primary and a secondary rainbow as well as some faint supernumerary bows.)

Scattering Effects

As pointed out in the last chapter, light will behave differently when scattered from small objects than when scattered from large ones. The effects of both types of scattering can be seen in the sky. The molecules of the air, itself, are smaller than the wavelengths of light, while water droplets and most dust particles are larger. Thus, the molecules produce wavelength-dependent scattering, while the water droplets that occur in clouds, for example, produce wavelength-independent scattering. This situation reveals why the sky is blue, clouds are white, and sunsets are red.

We will look first in detail at why the sky is blue. If you were on the moon or on any other planet that had no atmosphere, then the sky would be black, except when you looked directly at the sun. The presence of the atmosphere causes light to be scattered, and when we look in directions other than right at the sun, we see this scattered light. As we have mentioned, the molecules of the gases making up the atmosphere are smaller than the wavelengths of visible light. Consequently, shorter blue waves are scattered more strongly than the longer waves of other colors. Thus, we see this blue color when we look at a cloudless sky. We might expect the sun—when viewed directly—to appear redder than it actually is because of the blue light being preferentially scattered away. For the sun directly overhead, or nearly so, this preferential scattering of the blue light is not significant because the atmosphere is not thick enough to cause a large amount of scattering. However, when the sun is near the horizon, its light passes through more of the atmosphere. (Figure 14.5 shows these differences in the effective thickness of the atmosphere.) Thus, more blue light is scattered out of the way at dawn and dusk, giving us red sunrises and sunsets. This effect is intensified following the eruption of a volcano anywhere on the planet. Since the eruption spews dust into the atmosphere, the greater amount of wavelength-dependent scattering shows that the dust particles are very small.

The scattering of light also affects our view of scenes on the earth. As a patriotic song notes, distant mountains often have a purplish tinge. This tinge occurs even for tree-covered mountains that are very green at close range. Between the distant mountains and the observer, light is scattered by the air molecules, adding blue to that light coming from the mountains. The preferential scattering of short wavelengths adds this blue light, giving the mountains their purplish tint.

Another observable effect of scattering occurs on a foggy or misty day or when there is a lot of air pollution present. In this case, there will be wavelength-independent scattering from the large particles suspended in the atmosphere. White light scattered in this way remains white, and when it is added to the other colors from the scene being viewed, it causes those colors to become less saturated. Thus, on such a day, the colors in a scene will become very washed out.

When you are asked to imagine a beautiful sky, I am sure one picture that comes to mind is of white fluffy clouds against an azure background. We already have explained how scattering produces the blue sky, but why do the clouds appear white? Well, we see the clouds because light is scattered from them. However, the particles of water vapor that make up the clouds are larger than the wavelengths of visible light. Thus, all colors of light are scattered equally well, and the scattered light has the same color as the incident light. Clouds that are white are illuminated by the full spectrum of light coming directly from the sun. Sometimes, we see thick, heavy clouds that are very dark along their bottom edges. The clouds appear much darker in these areas because a much smaller amount of light reaches these areas to be scattered; the areas are in the shadow of other parts of the cloud. Clouds also can be illuminated by colored light. At sunrise or sunset, for example, the red sun can illuminate a cloud, and the cloud will appear to be red.

As well as affecting our view in the atmosphere, scattering can influence what we observe in the water. If we look into very clear water, we can see to great depths. The water will appear bluish in color for two reasons. First, the water often is illuminated with the bluish light coming from the sky. A second, but much less important reason, is that preferential scattering of light also occurs from the molecules of the water. Thus, when white light shines on the water, the blue is scattered more than the red. If very fine algae or particles of silt are dispersed throughout the water, however, the water will take on the color of the incident light. As with clouds, the fact that the scattered light takes on the color of the incident light means that those objects are larger than the wavelength of visible light. Of course, if the particles in either air or water get too large, they exhibit their own color, just as a red sweater does. Then, the light we see is being reflected from the objects, rather than being scattered by them. Examples of this situation include orange smoke rising from a manufacturing plant's stack and brown water in a stream following a heavy rain.

There is another effect we see in both air and water that is traceable to scattering effects: It is impossible to see shadows in either case unless there is suspended material. The exact opposite situation is also true. That is, a light beam shining through clear air or clear water cannot be detected from the side. The reason for this phenomenon is the same for both the light beam and the shadow: The molecules of air or water by themselves do not provide sufficient scattering to show where the light is shining. So, for example, if you are sitting by a clear pond or gently flowing stream on a bright sunny day, you will not be able to see the shadow of a tree on the water. If the water is shallow enough, you will see the shadow on the bottom, but not in the water, itself. Several days later, you might go to the same spot and find the water very turbid. (This change might happen, for example, if there had been a rain storm between your two visits to the spot.) In this case, the shadow of the tree will appear on the surface of the water. The particles of silt suspended in the water actually provide the surface on which the shadow falls. Even if the water were filled with ink so that it appeared very black, there would be no shadow. There is no shadow because the particles of ink absorb the light, so no significant scattering occurs.

In the same way, a light beam is visible only when there is sufficient material suspended in either the air or water to scatter the light. Thus, a visible beam of sunlight streaming in your window means the room is dusty! Water vapor in the air also can make sun beams visible. This phenomenon often happens on a warm, rainy day. A break in the clouds allows the sun through, and water vapor rising from the wet ground gives an ethereal picture.

Halos and Coronas

Under certain atmospheric conditions, you can look at a distant light source and see rings around the light. The source of the light might be the sun, the moon, or a distant street light. Two types of rings can form; they have different appearances, and they are created by different wave effects. A *corona* is generally a broad band of light circling the source and having a size (diameter) that is two or three times bigger than the source, itself.

A typical corona is sketched in Figure 14.6. The coloration across the ring is the clue we need to identify the wave effect producing the corona. Since red appears on the outer edge of the ring and blue on the inner edge, we can see (using the diagram in Figure 14.7) that the red light is bent more than the blue light. Further, from our discussion in the last chapter, we know that diffraction bends red light more than blue light. Therefore, the corona must be formed by diffraction. This is indeed the case, and usually, small water droplets in the air between the observer and the source of light cause the diffraction. The size of the corona is not predetermined; it depends upon the size of the water droplets that produce it. Just as with slits, smaller water droplets will produce more diffraction, so the corona will be larger. Also, if the water droplets are very uniform in size and are very evenly dispersed, the corona will be much more distinct. There is, however, a very narrow range of drop size that will cause sufficient diffraction to produce a corona.

It is easy to observe a corona with the following experiment. In a dark room or at night, breathe on a piece of glass (such as a microscope slide or a window) in such a way that a thin film of moisture is deposited on the glass. Then, look through the film at a small light source. (This source could be a distant street light or a flashlight bulb across the room.) As you look at this light source, you will see a corona surrounding it. On a dry day, the moisture will evaporate rapidly, and the corona will be short-lived. Even so, it is possible to see the coloration at the corona's inner and outer edges.

In contrast to a corona, a *halo* has a very definite size. Halos are produced by the refraction of light as it passes through ice crystals in the atmosphere. The shape of these ice crystals determines the amount of refraction that takes place. Since all ice crystals have the same shape, we always get the same amount of refraction. The structure of a water molecule (that is, the way the hydrogen and oxygen atoms are arranged in the molecule)

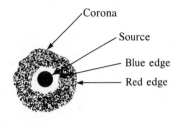

Figure 14.6. A sketch of a corona around a light source

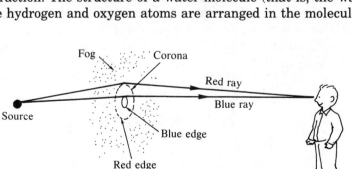

Figure 14.7. The rays of light producing a corona (Red is bent more than blue.)

A. An ice crystal

B. A snowflake

Figure 14.8. The hexagonal shape of solid water crystals

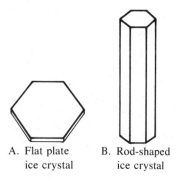

A. Flat plate   B. Rod-shaped
   ice crystal   ice crystal

Figure 14.9. Ice crystal shapes

Figure 14.10. The refraction of light by rod-shaped ice crystals

dictates an hexagonal symmetry for ice crystals. This shape is shown in Figure 14.8A, and we can see the same symmetry in a snow flake, as pictured in Figure 14.8B.

In the direction that is perpendicular to this hexagonal shape, the snow flake is very thin, but an ice crystal can be any thickness. Thus, the ice crystal can be shaped either like a flat, six-sided plate or a long, six-sided rod. These two extreme cases are shown in Figure 14.9. Now, when the atmosphere is filled with the rod-shaped ice crystals, we can observe a halo around a bright light source. Figure 14.10 shows how light from the source is refracted by the rod-shaped ice crystals to produce the ring around the source.

The ice crystals are distributed in the atmosphere in a very random way. Since all the ice crystals have the same shape, however, refracted light will reach the observer only from those crystals that are lined up properly. This situation is shown in Figure 14.11, where we see light coming in parallel rays from the source and going out in many different directions following refraction and reflection. Notice that many different orientations of ice crystals are present in this drawing. In spite of these many different directions, a number of light waves are heading off in the direction marked toward the observer. This change in direction of the light beam is what produces the circular halo around the light source. Because the ice crystals refract the light only in certain directions, the angle between the halo and the light coming directly from the source will have only certain values. The most common halo occurs at an angle of 22°; this angle is illustrated in Figure 14.12. Sometimes, we also might see a halo with an angle of 46°. This halo, of course, will form a much larger circle around the light source.

Just as with other refraction effects, there is also some dispersion present in the formation of a halo. The blue light is bent more than the red light, so a halo will have a blue edge on the outside and a red edge on the inside. Thus, there are two major differences between the appearance of a corona and that of a halo. The corona is a wide band located near to the light coming

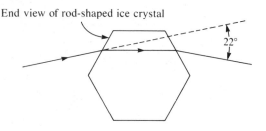

End view of rod-shaped ice crystal

22°

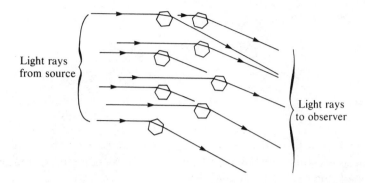

Light rays
from source

Light rays
to observer

Figure 14.11. Even though the orientation of ice crystals is random, light rays will move toward the observer in a preferred direction

directly from the source while the halo is a narrow ring located farther out. The second difference is the reversal of color order in the rings. (Both of these effects are shown in separate photographs in the color section of this text.)

Sun Pillars, Sun Dogs, and Sub Suns

A *sun pillar* is a column of light rising above the sun; a sun pillar occurs early in the morning or late in the evening when the sun is low in the sky, that is, near the horizon. *Sun dogs,* or as they are sometimes called, *mock suns,* also occur when the sun is low in the sky. (This effect also is termed a *parhelion.*) A sun dog is a bright spot in the sky to one side of the sun, but there are often a pair of spots, one on each side. Sun dogs are formed by the refraction of light through ice crystals, as happened with the halo. The distribution of the ice crystals for the sun dog is much more limited than for the halo, however. As shown in Figure 14.13, the ice crystals are trapped in a relatively thin layer of the atmosphere when sun dogs are formed. Thus, it is the same refraction that produces the halo, but here, the refraction occurs only on a level with the sun, and we see bright spots on either side of the sun rather than the full circle of the halo.

Although a sun pillar also requires the rod-shaped ice crystals to be present in the atmosphere, it is not produced by refrac-

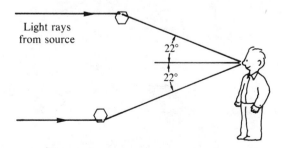

Light rays
from source

22°

22°

Figure 14.12. Refraction of light by ice crystals producing the 22° halo

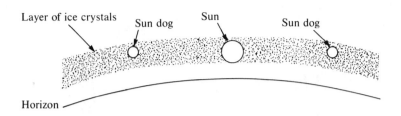

Figure 14.13. A layer of ice crystals producing sun dogs

tion of the light. Instead, the light is reflected off the faces of these crystals, as shown in Figure 14.14. As depicted in Figure 14.14, the rod-shaped ice crystals all lie horizontally in the atmosphere. However, the orientation of the hexagon around its axis is purely random. Thus, reflection occurs at any height but does so only along the line vertically above the sun. The reason ice crystals stay parallel to the ground is because they are heavy enough to be falling, and a long, thin, pencil-shaped object remains horizontal as it falls.

A *sub sun* sometimes can be seen when looking down from a high-flying aircraft. This phenomenon is another case where an image of the sun is produced by reflection of the light off ice crystals in the atmosphere. In this case, however, the crystals are wide and flat, rather than pencil-shaped. (See Figure 14.15.) These crystals also occur in a layer and are falling. As they fall, they tend to remain horizontal, much as a falling piece of paper would. This layer of flat ice crystals forms a crude mirror floating in the air. When a person is above this mirror, he can look down and see the reflection of the sun. As noted above, this image is named a sub sun.

Figure 14.14. Ice crystals producing a sun pillar by reflection of light

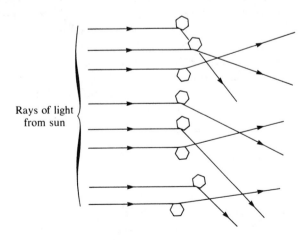

Rays of light from sun

Other Atmospheric Phenomena

Sometimes, in the late evening, after the sun has set and the sky has turned black, you may see some very white clouds against the black sky. These are called *noctilucent clouds* (which means clouds that are lighted at night). These clouds are so bright against the dark sky because they still are lighted by the sun. They are located very high in the atmosphere (at a height of 50 miles or 80 kilometers). Consequently, there will be a period of time in the evening when the clouds we see during the daytime are hidden by the Earth's shadow, but these very high clouds are still in sunlight. A diagram of this situation occurs in Figure 14.16. These clouds are so high in the atmosphere because they are composed of water drops that form on particles of cosmic dust. Ordinary clouds are much lower in the atmosphere and pass into the Earth's shadow soon after the sun sets.

We often associate the occurrence of a *mirage* with the desert. This association is only natural because stories about mirages there have much greater impact; someone who thinks the mirage of an oasis is an actual oasis makes an extremely critical mistake. Mirages, however, are not limited to the desert; they can occur anywhere. We often see a mirage when driving on a blacktopped road in the summertime: A puddle of water appears in front of us. But then, as we drive through the apparent puddle, there is no water. In this case, we have seen light from the sky that appears to come from the road's surface, and we interpret this appearance as light being reflected off the surface of a puddle of water. Instead of reflection, however, the light changes direction because of refraction. On a hot summer day,

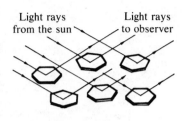

Figure 14.15. Flat ice crystals forming a crude mirror for creating a sub sun

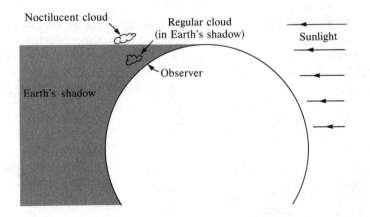

Figure 14.16. The lighting of noctilucent clouds

the air immediately above the road's surface becomes very hot. Light travels faster in the hot air than it does in the cool air above it, and this change in speed causes the light rays to bend. The path of the light rays for this simple mirage is illustrated in Figure 14.17.

All mirages are formed by the refraction of light waves in the atmosphere. Sometimes, the path of the light rays through the air is quite complicated. In general, though, the light travels faster through hot air than through cold air. Since the sunlight passing through the air heats the air very little, the surface of the land (or water) below the air is primarily responsible for the air's temperature. Surfaces such as blacktop roads or deserts are very hot, so the air above them is hot. In contrast, forests, grassy fields, oases, and bodies of water are much cooler. The cool air near these topographic features can cause the light to bend in the opposite direction to that shown in Figure 14.17. This bending can allow us to see objects that are normally out of sight because they are hidden by the curvature of the earth. When ships that are beyond the horizon still can be seen (because of a mirage), they seem closer than they actually are.

Another spectacular—although very transient—event occurs (under proper atmospheric conditions) just as the sun is setting. In the last view we have of the sun, the sun will be green. This *green flash* will last for only a few seconds, but it is so startling that it is quite noticeable. Several different processes must combine to produce this effect. First, of course, the blue part of the light coming from the sun must be scattered away by the air molecules. Then, there must be water vapor in the air because it absorbs orange and yellow parts of the sunlight. This absorption leaves only red and green. These two colors are separated by dispersion as the light is refracted in the air. As a result, there are two discs of sun in the sky—one green and one red—that almost are superimposed. Where they overlap, the sun appears yellow. However, the greater bending of the green light means we still will be able to see it after the red disc has disappeared behind the horizon. But we see the green flash for only the few seconds between the disappearance of the red disc and that of the green one.

Figure 14.17. The path of light producing a mirage for the driver of a car

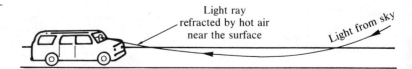

Light ray
refracted by hot air
near the surface

Light from sky

All of the atmospheric events we have discussed can be seen many places on the surface of the earth. By being aware of these phenomena, we may see them more often. But certainly, when we do see them, we will recognize that physical processes exist that explain these wondrous events.

Summary

There are many different atmospheric effects caused by light that have awed man through the ages. Myths and legends attempting to explain these effects are plentiful. In this chapter, we have discussed physical explanations for many of these atmospheric phenomena using our knowledge of waves. We saw how refraction (with dispersion) produces rainbows, halos, and sun dogs; how reflection can create sun pillars and sub suns; how diffraction produces coronas; how interference creates supernumerary arcs; and how scattering gives us red sunsets, white clouds, and blue skies.

Suggested Readings

The book *The Nature of Light and Colour in the Open Air,* by M. Minnaert (New York: Dover Publications, Inc., 1954) discusses the atmospheric effects we have covered. In addition, it describes many other effects. The reading level is certainly well within the reach of readers of this book, and the style is very readable.

The colors in the sky due to scattering effects, and the green flash are described more fully (and with a number of beautiful photographs of these effects) in *Sunsets, Twilights, and Evening Skies,* by Aden and Marjorie Meinel (New York: Cambridge University Press, 1983).

All of the atmospheric effects we have discussed are covered in greater detail (but without any greater mathematical sophistication) in the book *Rainbows, Halos, and Glories,* by Robert Greenler (New York: Cambridge University Press, 1980). The book contains many superb photographs of these effects and very nice diagrams to help explain the effects.

The column "Simple Experiments in Atmospheric Physics," which appears in the magazine *Weatherwise,* is usually interesting and often deals with topics described in this chapter. For example, the green flash is discussed in the December 1982 issue; scattering is discussed in the June 1983 issue; and a closely related topic, colors of the sea, is discussed in the October and December 1983 issues.

Other articles in *Weatherwise* that deserve special mention are "Chasing Rainbows," by Alistair Fraser (December 1983), and "What Causes a Double Rainbow?" by Thomas Schlatter (June 1983). The articles in *Weatherwise* are at the same level of reading sophistication used with this book.

There are a number of *Scientific American* articles that describe atmospheric phenomena. Eight of these articles have been reprinted in the book *Light from the Sky* (San Francisco: W. H. Freeman and Company, 1980), and four of these discuss topics that we have covered in this book.

These latter articles include "Mirages," by Alistair B. Fraser and William H. Mach (January 1976); "Atmospheric Halos," by David K. Lynch (April 1978); "The Green Flash," by D. J. K. O'Connell, S. J. (January 1960); and "The Theory of the Rainbow," by H. Moyses Nussenzveig (April 1977). As with all *Scientific American* articles, these articles have excellent diagrams and pictures. Most of the articles are at a slightly higher level of sophistication than that used in this book, and the article on the rainbow is quite technical. The other articles that appear in the book cover new effects: the aurora, the airglow, ball lightning, and the glory.

The article "Rainbows, Whirlpools, and Clouds," by Hans Christian von Baeyer, in *The Sciences* (July/August 1984), includes nice, nontechnical discussions of the rainbow and why the sky is blue and sunsets are red.

Methods for studying rainbows and supernumerary arcs in the laboratory (or at home) are described in "The Amateur Scientist" column in *Scientific American* (June 1980). The explanation of both the first and second order rainbows and supernumerary arcs is similar to that offered in this book. The diagrams are excellent, and there are several nice pictures.

The shapes of snow and ice crystals are discussed in "Snow Crystals," by Charles and Nancy Knight (*Scientific American*, January 1973). The primary thrust of the article is how the shape depends on the atmospheric conditions that occur while the crystal is being formed. This article probably provides more technical information than most readers would want. There are, however, some beautiful photographs of snow crystals that illustrate the wide variety of shapes that occur, even though we can see a hexagon in almost all of the crystal shapes.

Review Questions

1. Although they are not the same size, we can compare the primary rainbow with the most common form of halo. We find

a. the rainbow has a larger diameter than the halo, and both have the same order of colors.

b. the rainbow has a larger diameter than the halo, and they have their colors in the opposite order.

c. the rainbow has a smaller diameter than the halo, and both have the same order of colors.

d. the rainbow has a smaller diameter than the halo, and they have their colors in the opposite order.

2. Choose the correct statement from the following:

a. Since a rainbow and a halo are both created by refraction, they each have the same colors on the inside.

b. To observe a halo or a corona around the sun, you face toward the sun, while you face away from the sun to observe a rainbow.

c. Wavelength-dependent scattering produces halos, while wavelength-independent scattering produces coronas.

d. Rainbows can be seen only in the late afternoon because the sun travels through a thicker part of the atmosphere.

e. Noctilucent clouds are produced by diffraction of light by small ice crystals.

3. Sometimes we can see red clouds at sunset. This phenomenon occurs because

a. the clouds have red water droplets in them.

b. the clouds are illuminated with red light.

c. the clouds absorb blue and green light.

d. clouds in the afternoon are composed of smaller water droplets, so blue light is scattered better than red light.

e. we get a complementary effect to the green flash.

4. Distant mountains appear bluer than near ones. This phenomenon occurs because

a. the atmosphere absorbs red light.

b. distant mountains form a smaller image on the retina.

c. light from distant mountains is polarized.

d. light from nearby mountains is polarized.

e. blue light is scattered toward the eye by the atmosphere.

5. The primary and secondary rainbows differ in that

a. one reflection occurs inside each raindrop for the primary bow, and two occur for the secondary bow.

b. blue is on the outside of the primary bow and on the inside of the secondary bow.

c. the primary bow is created by diffraction, and the secondary bow is created by interference.

d. one refraction occurs inside each raindrop for the primary bow, and two occur for the secondary bow.

e. primary bows occur in the afternoon, and secondary bows occur in the morning.

6. Sun dogs and sub suns are

a. both caused by refraction.

b. caused by diffraction and interference, respectively.

c. caused by reflection and diffraction, respectively.

d. caused by refraction and reflection, respectively.

e. caused by diffraction and refraction, respectively.

15

Lasers and Holography

Important Concepts

- Light from a laser is produced by stimulated emission of radiation from prepared atoms.

- Laser light has three unique properties: It is monochromatic, highly collimated, and coherent.

- Holograms are permanent records of interference patterns.

- Using laser light with a hologram will give a true three-dimensional image of the original object.

In this chapter, we will study two relatively recent developments in physics that deal with waves. A laser is a new type of light source; the light it produces has some special properties. We will look at why these properties allow lasers to be used for tasks that other sources of light cannot perform. One of these unique uses for lasers is in holography, which is a new way of storing information. Our background in waves will be very helpful during our study of both these subjects.

Lasers as Light Sources

A laser is similar to a gas discharge tube in two ways. First, a laser converts electrical energy into light directly, that is, without heating some solid material. In the second resemblance, the laser is even more restrictive than the discharge tube. Whereas a gas discharge tube will emit a limited number of frequencies, the laser emits light of only one frequency. As with all light sources, the actual production of light occurs within atoms, and a complete understanding of the way laser light gets its unique properties requires knowledge about the atoms, themselves. To circumvent this need for understanding how atoms produce light, we will describe a mechanical model for producing waves on a string that has some similarities to the way a laser produces light waves.

The word *LASER* is an acronym. The letters stand for Light Amplification by Stimulated Emission of Radiation. The basic concept of the laser is that we prepare individual atoms within the device so that they are ready to emit radiation in the form of light waves. These atoms wait, however, until they are stimulated by other light waves before they actually emit their radiation.

We now will attempt to illustrate this concept using vibrating spring-mass systems and waves on a string, as shown in Figure 15.1. In picture A, the string is at rest in its equilibrium position, as are all the spring-mass systems shown. Each mass just touches the string when both are in their respective equilibrium positions. The masses are not attached to the string; they merely are touching it. For this model, these spring-mass systems are analogous to the atoms in a laser. When the mass vibrates, it will create waves on the string, just as the atoms create light waves. We also must make sure that the vibration frequency of each spring-mass system is the same. This requirement is necessary because we are trying to make this model behave like a laser, where all the atoms produce light waves of the same frequency.

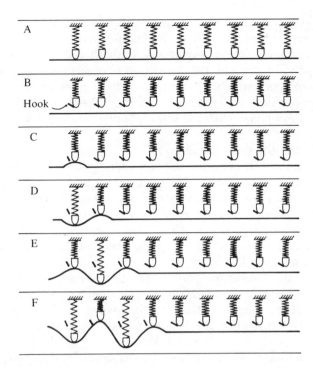

Figure 15.1. Amplification of a wave on a string by "stimulated emission"

A spring-mass system resting in its equilibrium position has, of course, no way of starting a wave on the string. Thus, we must "prepare" each system by giving it some energy. But we do not want any of these systems to vibrate yet. Thus, we push each mass up to the top and use a little hook to hold the mass there, as shown in picture B. In this way, we have stored some energy in each spring-mass system that can be delivered to a wave on the string whenever the hook is released. The hook we are using is special, however, in that it will release the mass whenever the mass is given a little upward push. And, of course, whenever a hook is released, a wave with the frequency of the spring-mass system will be sent along the string.

The unique feature of this complicated mechanical machine is shown in pictures C–F. As an initial pulse moves along the string, each mass, in turn, is raised slightly. This raising releases the hook holding the mass, allowing the mass to start vibrating. Thus, in picture C, the hook on the first mass has just opened. The hook for the second mass has just opened in picture D, and the first mass has pushed the string back down to its lowest position. This process continues through the remaining pictures, with each succeeding mass adding more energy to the wave. (Note that the amplitude of the wave as it moves along the

A typical laboratory-grade laser (Photograph provided courtesy of Metrologic Instruments, Inc., Bellmawr, NJ)

string is increasing.) But—and this is very important to recognize—each vibrating mass oscillates *in phase* with the original pulse. There is another way of describing this important consideration that will make the transition to describing light waves easier. We say that each vibrating mass puts waves on the string, but this activity is not a random process. Instead, all the waves are in phase with the original pulse, and, therefore, with each other.

We use the term *stimulation* to describe this process of an original pulse releasing the hook on a spring-mass system. Thus, each spring-mass system starts vibrating (and creating a new wave) when it is stimulated by the original wave. And each system waits—ready to emit a wave—until it is stimulated.

As we mentioned previously, this situation is analogous to what happens in a laser. Each spring-mass oscillator acts like an atom in a laser. The atom is energized or *pumped up* when it absorbs other electromagnetic radiation. Then, a light wave passes by and stimulates the atom. The atom emits another light wave that is in phase with the original wave. As more and more atoms are stimulated, the intensity of the light wave is amplified; thus, we have the name LASER.

Let's look in more detail at the construction of a typical gas laser. The diagram in Figure 15.2 shows the essential parts of such a laser. The tube is filled with a gas having the correct atomic properties and is sealed. The two electrodes allow the introduction of electrical energy into the tube so that the gas atoms can be pumped up. At each end of the tube, there is a mirror. These two mirrors must be aligned carefully so that they are perpendicular to the tube. Then, light originating in the tube will reflect back and forth between these mirrors many times. (The need for the mirrors will be described shortly). One of these mirrors is designed so that not all the light reaching it is reflected; some light passes through. We characterize such a mirror as *partially reflecting*. The light leaking through this mirror is the laser beam.

The stimulated emission of radiation that occurs inside a laser is not as simple as the stimulated emission of waves in the mechanical model described earlier. A fundamental complication occurs because light waves can be emitted in any direction; whereas, the wave on the string can be sent off only in two directions. However, when an atom is stimulated by a light wave to emit radiation, the new radiation is not only in phase with the incident light wave, but it also moves in the same direction. By having mirrors at each end of the laser tube, light waves that are

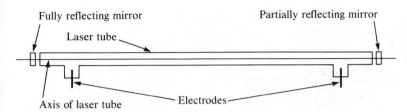

Figure 15.2. The essentials of a gas laser

moving along the tube are reflected back and forth. Thus, these light waves have many opportunities to stimulate additional atoms. In contrast, any light waves that are emitted at an angle will escape out the side of the tube. As a consequence, the light traveling along the tube gets amplified and also travels back and forth many times before it escapes through the partially reflecting mirror.

Because of the lasing action and its many reflections inside the laser, the light emerging from the laser has three unique properties. First, it is *monochromatic*. By this term, we mean the light has only one color, that is, only one wavelength or frequency. (Actually, there is a very small range of frequencies or wavelengths in the light from a laser. The spread is so small, however, that we will neglect it and will talk in terms of only one wavelength or frequency.) The second unique property of laser light is that it is very highly *collimated*. This term means the size of the beam does not increase very much. Since the light in a laser bounces back and forth between the mirrors many times before emerging, it must be traveling on a path that is parallel to the tube. If its path makes even a small angle with the axis of the tube, it eventually will miss one of the mirrors and be lost. This situation is shown in Figure 15.3, where the light is almost parallel to the axis; the light makes five transits of the tube length before missing a mirror. While light from other sources can be focused into a collimated beam using lenses or curved mirrors, the degree of collimation is not as great as for a laser. The fact that light from other sources contains many wavelengths contributes to the difficulty in achieving very high collimation. The third unique property of laser light is its *coherence*. This coherence actually takes two forms: *time coherence* and *spatial coher-*

Figure 15.3. The fate of a ray of light that starts off nonparallel to the laser tube axis

The needle in this picture is lying on a piece of ⅛-inch-thick steel. The small hole next to the needle was produced with a laser. (Photograph provided courtesy of Hughes Aircraft Company.)

ence. Time coherence means the frequency of the light stays the same over long periods of time, compared with the period of the wave. Since visible light waves have periods that are much less than a trillionth of a second, the long periods of time we are referring to may be only a second or so. Spatial coherence means that all the waves in the laser beam are in phase with each other.

These properties of laser light are very different from those for light produced by other sources. In every other light source, light waves are emitted in a very random fashion. Therefore, we can neither predict nor control the direction, frequency, or phase of individual light waves.

Uses of Laser Light

The unique properties of laser light just mentioned allow it to be used for tasks that other types of light cannot perform. In each of the cases we discuss here, we will emphasize which properties of laser light are being utilized in that application.

The monochromatic and collimation features mean a laser beam can be focused to a much smaller spot than ordinary light. When this is done, the entire power contained in the beam is concentrated on that small area. The temperature of the spot rises to a very high value; sometimes (for high-powered lasers), the temperature exceeds that found on the surface of the sun.

Such high temperatures make the laser an excellent device for cutting a wide variety of materials, from metals to animal tissue. In the former case, the laser works much like a cutting torch. The metal is heated well above its melting temperature over a very small region. As the laser moves, it makes a thin cut in the metal. (The thickness of the cut is the width of the beam.) The laser is better for this purpose than a gas torch because the laser's cut is thinner and smoother.

When a surgeon uses a laser in place of a scalpel, there are also certain advantages. The very high temperature gives a cauterized cut, so there is much less bleeding than with the usual surgical techniques. Focusing the laser light beam causes the small spot to occur at a particular distance from the lens. (See Figure 15.4.) Thus, the depth of cut by a laser scalpel also can be controlled very easily. Another advantage occurs in eye surgery where a wide, diffuse laser beam can be sent through the lens and focused on the retina. This technique is used for at least two types of retinal problems. Hemorrhaging of blood vessels (that is, the breaking of blood vessel walls) in the retina is a leading cause of blindness. Potential problems can be corrected by burning the weak walls of blood vessels using a focused laser beam. The resulting scar tissue gives added strength that prevents rupture. A detached retina also can be corrected by using laser burns to "spot-weld" the retina back onto the rear of the eyeball. In both these operations, there is a great advantage in using a laser because the laser acts directly on the retina without disturbing any other parts of the eye.

In an entirely different type of application, the laser can be used to carry information from one place to another. In this application, the laser performs the same function as a much longer wavelength radio wave. The essential properties of laser light for this application are its coherence and monochromaticity. In both the cases of radio waves or laser light, information is sent by modifying (or modulating) the waves. When this modify-

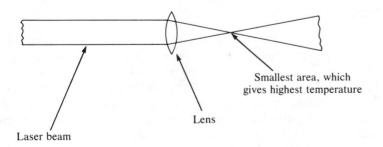

Figure 15.4. Focusing a laser beam

Smallest area, which gives highest temperature

Lens

Laser beam

ing is done, we change either the amplitude or the frequency of the wave. These processes are called *amplitude modulation* (AM) or *frequency modulation* (FM), respectively. Both types of signals require a coherent and monochromatic wave, so, for electromagnetic waves in the visible range of frequencies, only laser light will do. In one way, lasers are actually better than radio waves for communication. The rate at which we can send information on any wave depends on the wave's frequency. It is easiest to understand this statement by thinking in the following way: We basically can send one piece of information on each crest of the wave. For high frequency waves, there are more crests every second, so more information can be sent during a given period of time. Thus, with the use of lasers for communication, we are able to send far more information than when radio frequency waves are used. However, light waves cannot penetrate clouds or diffract around buildings nearly as well as radio waves. Thus, light pipes, such as those discussed in Chapter 11, carry the beams of laser light from one location to another, much as a wire carries an electrical signal for telephone conversations.

A third use for lasers is in the production of holograms. As we will see presently, a hologram is just a permanent record of an interference pattern. Without coherent waves, there would be no interference pattern; therefore, the coherence of waves in the laser beam is vital for holography.

Surveyors and construction engineers now use lasers for precise alignment and measurement. In this application, the highly collimated laser beam is a straight line through space. For this purpose, a laser is easier, faster, and more accurate than the transit it replaces.

The examples cited here are just a few of the many applications for which lasers are now being used. Since the first laser was built by T. H. Maiman in 1960, the variety of uses has spread dramatically.

Holography

A visual image of a subject can be recorded with a photograph. The image recorded on the film shows the view seen by the camera at the instant we take the picture. As we have seen earlier, light from the subject is focused onto the film by the camera lens. Thus, light from one point on the subject reaches one point on the film. (See Figure 11.24.) Photography is a method of storing the visual information contained in the image.

An entirely different method of storing visual information

occurs in holography. Just as in photography, we use film to permanently record the information. But that is the only similarity between the two processes. As mentioned earlier, a hologram is not an image of the recorded scene. Instead, it is a record of an intereference pattern between two beams of coherent light. We can see the original scene when we again use coherent light to illuminate the hologram. This process creates a virtual image, but this virtual image is different from the usual photographic image in that it is truly three-dimensional. If the recorded scene has both close and distant objects in it, then we can change our perspective when viewing the hologram and watch the farther objects disappear behind the nearby ones. To understand this and other unique characteristics of holograms, we must see how holograms are made.

The diagram in Figure 15.5 shows how the two light beams are used to create the holographic interference pattern. A single beam of light emerging from the laser is directed toward a partially reflecting mirror. Part of the beam passes straight through the mirror, and part of it is reflected; thus, this mirror serves as a *beam splitter*. Both of these beams that leave the mirror are expanded in size by lenses. One of the beams, called the *reference beam*, then falls directly onto the film that will become the hologram. The other beam, the *object beam*, illuminates the object. Some of the light carried by the object beam reflects off the object and also hits the film. This reflected object beam light interferes

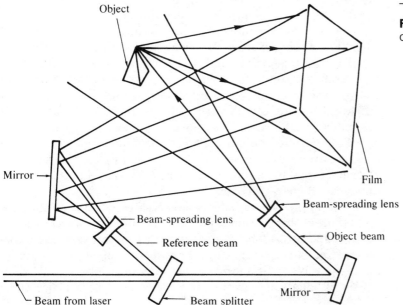

Figure 15.5. The production of a hologram

The arrangement of the apparatus for producing a hologram. The white box on the lower left is the laser; its beam moves across the bottom of the picture to the right. The beam first encounters the beam splitter and then (at the lower right) encounters a mirror. The film on which the interference pattern is recorded is at the right. The two cylindrically shaped objects near the bottom of the picture are the beam-spreading lenses; the one for the object beam is on the right. (Ronald R. Erickson, Museum of Holography)

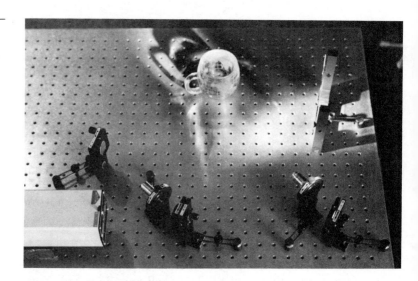

with the reference beam, producing an interference pattern. Points of constructive interference expose the film, while points of destructive interference do not. It is important to recognize that light from each point on the object reaches every point on the film. This situation is shown in Figure 15.5 for light being reflected from the top of the pyramid. Of course, light from every other point on the pyramid and from all the other objects in view does the same thing. As a result, the interference pattern on the hologram will be very complicated.

The film is developed in the usual way to produce the hologram, and we put the film back into the same location it occupied when it was being exposed. But, we remove the object so that the only light falling on the film is the reference beam. The recorded interference pattern on the hologram acts as a grating for each point in the original object. As shown in Figure 15.6, the rays diffracted by this complicated grating create an image of the original object. Since each section of the hologram is produced by object beam rays from all points in the scene, each section creates its own view of the scene. The different perspectives of each section of the hologram gives the hologram its true three-dimensional characteristic.

A different type of hologram is created when the reference and reflected object beams approach the film from opposite sides during exposure, as shown in Figure 15.7. In this case, the interference (or superposition) of the two sets of waves produces standing waves. (Standing waves always are produced when waves of identical wavelengths move in opposite directions

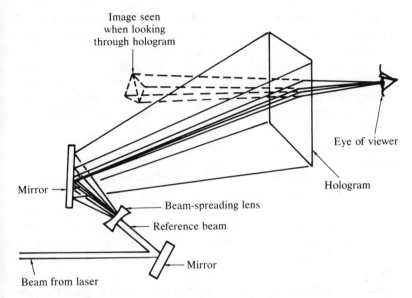

Figure 15.6. Viewing a transmission hologram

through each other!) The film is thick, so there are standing waves within it, and it is exposed where the antinodes are located. The antinodes, of course, are separated by one-half the wavelength of the light used. When the film is developed, the exposed portions act as partially reflecting mirrors. Since the exposed portions are so close together, they act like surfaces of thin films and create interference effects for light reflected from them. Consequently, such a hologram can be viewed with white

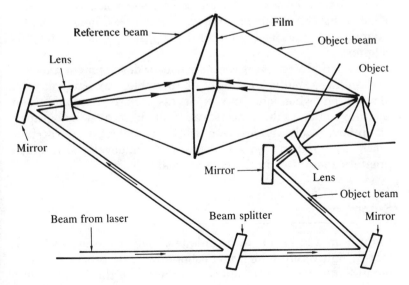

Figure 15.7. Making a "white light" hologram

light. We will see the image because thin film interference will give constructive interference for only one wavelength of light. It would be the same wavelength as the original laser light except that the film shrinks slightly during development, causing the reflecting surfaces to move closer together. Thus, light having a wavelength shorter than that of the laser gives constructive interference, and the image we see has a different color.

Since a hologram is just the photographic record of the interference between the reference and object beams, it is critical that this interference pattern be maintained while the film is being exposed. If any part of the apparatus moves during the exposing of the film, the interference pattern will be "washed out," and the hologram will be ruined because the film will only record the points of constructive interference where the light intensity is greatest. The points of destructive interference will not expose the film; no light energy is arriving at those points. But if the interference pattern changes so that constructive interference occurs where there was originally a point of destructive interference, then that place will be exposed, and no pattern will be recorded. (In order to have a pattern, there must be areas of both light and dark on the film.) It would not matter if the dark and light regions were interchanged on the film, however, because this would not change the pattern. Thus, a hologram does not have negative and positive images.

As we have seen, holography is the recording of an interference pattern between two beams of waves. We have described the process in terms of light, but the process works for any type of coherent, three-dimensional wave. Acoustical holograms, for example, are described in one of the suggested readings for this chapter.

Holography is a very recent scientific development and still is changing rapidly. Currently, it resembles the early days of photography when long-time exposures were required and when the range of suitable subjects was limited. Because holography has inherent advantages over regular photography for certain types of permanent records, it is sure to be developed to much greater capabilities in the coming years.

Summary

Light from a laser has three unique properties: It is monochromatic, highly collimated, and coherent. These properties arise because atoms within lasers store energy and then emit visible radiation when they are stimulated by existing light waves; thus, we get the acronym "laser" for light amplification by stimulated emission of radiation. Among their many applications, lasers can be used in communication, for cutting, as an alignment instrument, and for holography. A hologram is a permanent record of the interference pattern between a reference beam and an object beam. A true three-dimensional image of the original object can be reconstructed using the hologram and light from a laser.

Both lasers and holography are relatively new scientific developments. During the coming years, we will see many new applications for both. Because they are recent advances, we think of them as being quite sophisticated. Indeed, they are, but we can understand many of their unique features using the knowledge of waves we have developed.

Suggested Readings

Lasers and Holography, by Winston E. Kock (Garden City, N. Y.: Anchor Books/Doubleday, 1969), describes these topics at a level that could be understood easily by someone who has read this book.

One use for lasers is discussed in the article "Communication by Laser" (*Scientific American,* January 1966.) This article discusses some of the unique properties of laser light that make it ideal for a means of communication.

The production of holograms with another type of wave is discussed in the article "Acoustical Holography" (*Scientific American,* October 1969). The article describes how holograms are made on the surface of a tank of water using sound waves. The holographic interference pattern then is visualized by reflecting laser light off the water's surface.

Review Questions

1. Bluish green laser light is sent through a prism. Which of the following happens?

    a. The light is broken up into the additive primaries blue and green.

    b. The light is broken up into the subtractive primaries yellow and magenta.

    c. The light is not bent.

    d. The light is bent, and only bluish green is seen.

2. The property of laser light most important to the production of holograms is that laser light

   a. has a short wavelength.

   b. can be spread out to illuminate the entire object.

   c. can be reflected with a mirror.

   d. can be diffracted.

   e. is coherent.

3. A laser produces a beam of light with a wavelength of 500 nm and a frequency of 600 trillion Hz. Another laser produces a beam of light with a wavelength of 480 nm and a frequency of 615 trillion Hz. At a point where the beams pass through each other,

   a. beats will occur with a wavelength of 200 nm.

   b. standing waves will be observed.

   c. beats will occur with a frequency of 15 trillion Hz.

   d. no standing waves or beats are possible; standing waves or beats occur only for longitudinal waves.

4. The negative of a hologram produces

   a. a negative image.

   b. the same image as the positive hologram.

   c. an inverted and left-right reversed image.

   d. an image whose color is the complementary color of the original laser light.

   e. no recognizable image.

5. Long John Silver makes a *hologram* of his buried treasure map (the location of the treasure is on the left side of the *map*) and cuts the hologram in two pieces. Jim Hawkins steals the right-hand piece. Choose the best of the following conclusions.

   a. Jim wasted his effort—only the left side of the hologram will show where the treasure is buried.

   b. Jim was lucky to get the right-hand piece because the image on the hologram is reversed.

   c. Jim was smart. He knew that each half of the hologram can reproduce the entire map, so he did not bother looking for the other piece.

d.   Long John Silver was very sly. He knew that both pieces of the hologram were needed to reconstruct the map.

6.   It is possible to produce holograms with sound waves (although such sound wave holograms are not recorded on photographic film). This possibility exists because

a.   audible sound and visible light have about the same frequency and wavelength.

b.   laser light is not polarized.

c.   laser light is a longitudinal wave.

d.   coherent sound waves can be produced.

7.   A hologram is made using one laser for the reference beam and a different laser for the object beam. When we view the hologram, we see

a.   only a two-dimensional image of the object.

b.   a true three-dimensional image of the object.

c.   no image at all.

d.   two overlapping, three-dimensional images.

16

Epilogue

We now have completed the task we set out to accomplish in Chapter 1: We have looked at how waves can help us understand the world around us. In this chapter, we will review what we have discussed, and we will emphasize some important points. We also will take a quick and very cursory look at other examples of wave behavior in the realm of quantum physics. These concepts will be even more difficult to grasp than the topics we already have covered because these concepts are further from our everyday experiences. In spite of their esoteric nature, however, these concepts are just new applications of the same basic wave behaviors we have worked so hard to understand.

We started our discussion by looking at vibrational motion because waves transfer vibrations from one location to another. Simple objects such as tuning forks and pendula can only vibrate in a simple way, while extended objects such as strings, plates, or blocks of jello can vibrate in either simple or complicated ways. When an extended object vibrates in a simple way, it is vibrating in only one of its normal modes; any complicated motion can be described as a superposition of two or more normal mode vibrations. In either case, the oscillation of an extended object can be explained on the basis of waves moving in opposite directions through the object and superposing to form the perceived motion. In general, two waves of equal magnitude and frequency moving in opposite directions through a medium will produce a standing wave; all normal mode vibrations of extended objects are standing waves. We can visualize this process most easily when we use a string as the extended object; for this reason, we used strings to develop our understanding of standing waves. We saw how the normal modes of vibration that are present depend on the way the string is terminated at each end. The overtone structure of only odd harmonics for a string with unlike ends occurs because a pulse must make two round trips on the string before being superposed with another pulse. In contrast, a string that has ends that are the same can have all harmonics present because each pulse needs to make only one round trip before being superposed with a subsequent pulse.

The beauty of the study of waves is the wide applicability of the ideas. For example, the same concepts developed to explain standing waves on strings allow us to understand standing sound waves in organ pipes, seiches in bodies of water, and the notes produced by the bars in a xylophone. (We have not discussed this last example, but the ideas are the same, so you should have no trouble in applying them to this case.)

It is also true that all waves exhibit refraction, reflection, diffraction, and interference. Because light and sound exhibit all these phenomena, we conclude that there are light and sound waves, even though we cannot see the waves, themselves. Not only do all waves show these behaviors, but they also exhibit these behaviors in the same way. That is, for refraction occurring at an interface, the wave direction is always closer to the normal in the slower medium; the angle of reflection is equal to the angle of incidence for all waves; greater diffraction occurs through a narrow opening when the wavelength to opening size is larger; and constructive interference always occurs when two waves are in phase with each other.

Physicists have made great strides in their quest to understand nature by using the similarities that occur among various types of waves. The differences that exist among these waves are also important. The most apparent example is the use of polarization of light to recognize that light is a transverse wave. Historically, many physicists rejected the idea that light is a wave because they were thinking of longitudinal waves (from their experiences with sound) and could not reconcile the behavior of light with that of sound. Indeeed, Sir Isaac Newton (one of the greatest physicists who ever lived) did not believe in the wave theory of light. He believed that light sources emit a stream of particles, which he called corpuscles. He thought that these corpuscles then bounce off reflective surfaces and change speed to produce refraction when they move into a different medium. It is extremely difficult to explain the wave behaviors of diffraction and interference with such a corpuscular theory of light, however. Consequently, when these effects were discovered, most physicists accepted that light is a wave. The experimental determination that light travels slower in material such as glass and water than in air enhanced the belief in light waves because the explanation of refraction using corpuscles requires them to travel faster in media where the direction of light is closer to the normal. The final nail in the coffin of the corpuscular theory was the explanation of polarization effects using transverse waves.

Thus, it seems that scientists have been able to understand the behavior of light using the concepts of waves. But the story does not end here. As you may have noticed, we have not discussed the production of light waves in any detail; we have only said that they are produced at the atomic level. At about the beginning of this century, physicists resurrected the idea that light is a stream of particles in order to explain how light interacts with matter during production and detection or absorption processes. This time, however, the physicists used the two ways of looking at light in a complementary manner. Waves are used to explain the movement of light from one location to another, and particles are used to explain emission and absorption of light by matter.

This dual way of describing light is necessary because light is more complicated than sound waves or the waves on a string, or water, or on Chladni plates. The theory also requires the two ways of thinking about light to be connected, and this connection is a philosophical underpinning of the quantum theory. These particles of light were named *quanta* by Max Plank and *photons* by Albert Einstein, and we still use these names today. Thus, atoms emit photons of light, and the light behaves as a wave. This behavior is accomplished because the photons move according to the dictates of a wave. That is, there is a wave that tells the photons how to move. This wave is called a *probability wave* because it determines the probability that the photons of the light actually will be at a particular location. Thus, this wave is even more esoteric than a sound wave, and we cannot even see sound waves! However, this probability wave behaves—in many ways—just like the other waves we have looked at. Indeed, all the behavior we ascribed to light waves really is controlled by this probability wave.

Thus, we see that there is a duality when it comes to describing light: a particle picture whereby light is a stream of photons and a wave picture whereby light is a transverse wave. This duality is not limited to light! All subatomic entities such as electrons, protons, neutrons, and other objects we ordinarily

An example of how particles exhibit wave behavior. Picture A shows a diffraction pattern produced by X rays (electromagnetic waves) passing through a thin aluminum foil. Picture B shows the diffraction pattern produced by a beam of electrons passing through the same foil. (PSSC Physics, 2nd edition, 1965; D.C. Heath and Company with Education Development Center, Inc. Newton, MA.)

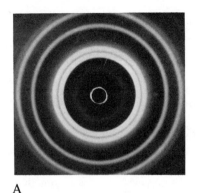

A

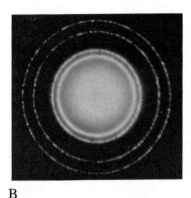

B

think of as particles also exhibit this dualism. For example, we can see diffraction and interference effects with these particles.

The investigation of these quantum effects is beyond the scope of our investigations. However, having digested the material on vibrations and waves in this book, we will find such further study easier. The probability waves of the quantum theory are yet another example of the concepts we have been discussing. Thus, it is indeed true that waves are vital in understanding the world around us.

Glossary

Amplitude the maximum displacement from the equilibrium position of a vibrating object

Antinode the location on an extended vibrating object where the amplitude of vibration is at its maximum

Attack the way a tone starts; the different attacks produced by various musical instruments are clues our ears use to distinguish the instrument playing the tone

Beat a vibration with a varying amplitude which is produced by combining (using the principle of superposition) two vibrations with slightly different frequencies

Brewster's Angle the angle of incidence for which the reflected beam is totally polarized

Characteristic Frequency see natural frequency

Chladni Plates flat metal plates that are vibrated by bowing and on which sand is sprinkled so that it will collect along nodal lines to show their locations

Coherence an unchanging phase relationship; two sources are coherent if the phase difference between them is always the same

Compression a place in a longitudinal wave where the medium is squeezed together; also called a condensation

Concave a curved interface which is depressed inward as viewed from the surrounding medium

Condensation see compression

Convex a curved interface which sticks out into the surrounding medium

Coupling the connection of one vibrating system to another

Crest that part of a transverse wave which is at its maximum upward displacement

Critical Angle the angle of incidence at which total internal reflection starts

Cycle the motion of an oscillator until it starts repeating itself

Cyclic Motion motion that repeats; that part of the motion within each repetition is termed a cycle

Damping Time the time it takes an oscillator's amplitude to decay from any value to one-half that value

Dashpot a mechanical device which can be used to absorb the energy carried by a wave on a string

Decay either the change of an oscillator's amplitude with time or the way a musical tone ends

Degree of Freedom an independent way a vibrator can move

Difference Tone a tone created by the superposition of two other tones; its frequency is the difference between the frequencies of the two tones producing it

Diffraction the bending of waves around edges of obstacles; longer waves bend (diffract) more than shorter waves

Dispersion the separation of waves of different wavelengths because of a difference in speeds

Doppler Effect the change in detected frequency which occurs when either the wave source or the detector (or both) are moving

Driver an oscillator which causes another oscillator to start vibrating

Driving (a string) starting the string vibrating, usually by bowing, hitting, or plucking

Equilibrium Position the place where a vibrating object comes to rest, usually at the midpoint of the oscillator's movement

Focal Point the point in space where rays passing through a converging lens meet or where those passing through a diverging lens appear to come from

Frequency the number of cycles an oscillator completes during one unit of time—usually during each second (see also Hertz)

Fundamental the normal mode with the lowest frequency (also used to refer to the frequency of this normal mode)

Harmonics normal modes that have frequencies which are integral multiples of the fundamental's frequency

Hue   the subjective quality for describing the color of light; most closely related to the dominant frequency in the light

Incandescent   hot enough to emit visible light

Incident   the wave (or ray) which is approaching an interface between two media

Integral   (such as Integral Multiples or Integral Number)   integer (whole number)

Intensity   the objective quality of a vibration which describes its energy; most closely related to the subjective quality of loudness

Interface   the boundary between two media where a wave has different speeds

Interference   the superposition of waves emanating from different sources

Laser   a source of monochromatic, coherent, and highly collimated light

Lens   a device used to focus waves

Lissajous Figure   the path followed by a point which vibrates in two perpendicular directions at the same time; the relative frequencies and phases of the two vibrations determines the shape of the figure

Longitudinal Wave   a wave in which the vibrations of the medium are in the same direction as (along) the movement of the wave

Loop   the distance between adjacent nodes of a standing wave; the name derives from the shape of the medium as it vibrates

Melde's Experiment   a tuning fork driving an attached string

Medium   the matter or material through which a wave moves

Monochromatic   light that has only one color; that is, a light of only one frequency

Natural Frequency   the frequency at which an oscillator vibrates when it is free to move by itself; when driven, an oscillator will respond better at this frequency than at any other

Node   a place on an extended vibrating object which does not move

Normal Mode   a mode of vibration for an extended object for which every piece of the object moves with the same frequency

Octave   fundamental division in a musical scale; moving up an octave doubles the frequency

Optical Density   the measure of how much a medium slows down light

Oscillator   a device which oscillates (vibrates)

Oscilloscope   an electronic instrument which allows us to see vibration patterns and Lissajous figures

Overtones   normal modes other than the fundamental

Period   the time during which an oscillator completes one cycle of its vibration

Periodic Motion   motion which repeats itself; see cyclic motion

Phase (or Phase Angle)   the variable which describes the location of an oscillator within a cycle of its vibration; the phase changes from 0 to 360 as the oscillator moves through each cycle

Pitch   the subjective quality describing relative high or low tones; related most closely to the objective quality of frequency

Plane Wave   a wave with wavefronts that are planes (in three dimensions) or straight lines (in two dimensions)

Polarization   a transverse wave is polarized when its vibrations are restricted to one direction only

Principle of Superposition   the fundamental idea that the net effect due to several separate ones is the sum of the individual ones

Rarefaction   the location in a longitudinal wave where the medium is spread apart

Ray   the path followed by a point moving along with a traveling wave

Real Image   an image that can be projected (and focused) on a screen

Reflection   the change in direction of waves because they bounce off an interface

Refraction   the change in direction of waves due to a change in their speed

Resonance   the situation where energy is readily transferred from one oscillator to another; that is, very good response

Resonant Frequency   the frequency at which resonance occurs; see also natural frequency and characteristic frequency

Response   the vibration which results from an oscillator being driven by another oscillator

Response Curve   the graphic depiction of how an oscillator responds at various driving frequencies

Saturation   the subjective quality describing

the purity of a light; most closely related to the mixture of wavelengths in the light

Scattering the change in direction of waves which encounter small obstacles; wavelength dependent scattering occurs when the size of the obstacles are smaller than all wavelengths in the incident waves

Seich a standing water wave

Sine Curve a mathematical curve which is the vibration pattern for a pure tone

Simple Oscillator an oscillator which can vibrate in one way only; so it has only one natural frequency and one degree of freedom

Speed the rate at which an object or a wave moves

Spherical Wave a wave having wavefronts shaped as spheres (in three dimensions) or circles (in two dimensions)

Standing wave a normal mode of vibration created by the superposition of two identical waves traveling in opposite directions

Tension the tightness of a string

Timbre see tonal quality

Tonal Quality the subjective quality describing how the tone sounds; related most closely to the objective quality of vibration recipe

Total Internal Reflection the situation which occurs when no refraction occurs; it happens only when a wave moves from a slow medium to a fast one and the angle of incidence is greater than the critical angle

Transverse perpendicular; see transverse wave

Transverse Wave a wave in which the motion of the medium is in a direction perpendicular to the direction of the wave motion

Traveling Wave a wave that moves from one location to another

Tremulo rapid variation of the amplitude of a vibration producing sound

Trough a depression of the medium within a traveling wave

Tuning Fork a device which acts as a simple oscillator; it produces a pure tone

Velocity another word for speed although a description of the motion's direction is usually included when the word *velocity* is used and such information is not included when the word *speed*

is used

Vibration a repetitious motion

Vibration Pattern a picture of how an oscillating object moves during its vibration

Vibration Recipe the frequencies and amplitudes of the simple vibrations which combine to produce a complex vibration

Vibrato sound having a rapid variation of pitch

Virtual Image an image which can be seen by looking into an optical element (lens or mirror) but which cannot be projected onto a screen

Wave the mechanism of moving energy from one location to another without moving any matter; the energy is moved as vibrations

Waveform another name for a vibration pattern

Wavefront a line or surface of constant phase of a wave; it is an easy way to show the shape of a two- or three-dimensional wave

Wavelength the distance between the same points in adjacent cycles of a traveling wave; a simple example is the distance between adjacent crests of a wave

Answers to End-of-Chapter Questions

Chapter 1   1 c, 2 a, 3 a, 4 c, 5 e, 6 b

Chapter 2   1 b, 2 b, 3 b, 4 e, 5 c, 6 c, 7 c, 8 d, 9 c, 10 e, 11 d

Chapter 3   1 c, 2 c, 3 d, 4 b, 5 e, 6 a, 7 c, 8 e, 9 b, 10 c, 11 d

Chapter 4   1 b, 2 c, 3 d, 4 c, 5 a, 6 c, 7 a, 8 b, 9 c, 10 e

Chapter 5   1 a, 2 b, 3 b, 4 b, 5 d, 6 a, 7 c, 8 e, 9 b

Chapter 6   1 c, 2 c, 3 a, 4 a, 5 b, 6 b, 7 d, 8 e, 9 b, 10 e, 11 b

Chapter 7   1 b, 2 d, 3 d, 4 a, 5A c, 5B d, 6 c, 7 d, 8 e, 9 b, 10 d

Chapter 8   1 e, 2 a, 3 a, 4 b, 5 b, 6 c, 7 c, 8 c, 9 d, 10 b

Chapter 9   1 b, 2 d, 3 d, 4 c, 5 b, 6 a, 7 e, 8 a

Chapter 10   1 c, 2 e, 3 d, 4 c, 5 a, 6 e, 7 a, 8 c, 9 c, 10 e, 11 d, 12 a, 13 a

Chapter 11   1 d, 2 d, 3 a, 4 b, 5 C, 6 G, 7 B, 8 E, 9 c, 10 A, D, F, & H 11, B, C, E, & G, 12 A, D, E, & G, 13 B, C, F, & H, 14 b, 15 c

Chapter 12   1 c, 2 c, 3 c, 4 d, 5 b, 6 a, 7 b, 8 a

Chapter 13   1 a, 2 d, 3 a, 4 d, 5 b, 6 b, 7 d, 8 d, 9 d, 10 d, 11 e

Chapter 14   1 b, 2 b, 3 b, 4 e, 5 a, 6 d

Chapter 15   1 d, 2 e, 3 c, 4 b, 5 c, 6 d, 7 c

Index

Alexander's dark band, 303
AM (Amplitude Modulation), 324
Amplitude, 19, 21, 26, 93
Analyzer (Polarization), 286
Angle of incidence, 215, 217
Angle of reflection, 217
Angle of refraction, 215
Antinodes, 49, 154, 165
Antiphase, 24
Antisolar point, 301
Attack, 76, 83, 92
Auditory canal, 98

Basilar membrane, 98, 100
Bassoon, 183
Beam splitter, 325
Beats, 34–36
    Frequency, 95
    Surf beat, 134
Birefringent crystals, 290
Blue sky, 304
Bore, 183
Bowing a string, 55, 179, 181–82
Breaking of waves, 135–36
Brewster's angle, 290
Bridge, 72, 179
Brightness, 250
Bugle, 187

Cello, 179
Characteristic frequency, 25
Chladni plate, 59, 167
Chromaticity diagram, 252
Clarinet, 183
Coherence, 275, 321
Collimation, 321

Color, 198–202
    Addition, 199 and Plate 3
    Subtraction, 199 and Plate 4
    of Objects, 201
    Printing, 201
Complementary colors, 200
Condensation, 111
Consonant tones, 78
Constructive interference, 272–81,
    303
Continuous string, 48
Corona, 306
Corpuscles, 334
Coupling, 28, 180
Crest, 107
Critical angle, 219
Cycle, 19

Damping, 24, 58
Damping time, 25, 28, 55, 58
Dashpot, 149–51
Decay, 83, 92
Degree of freedom, 45
Destructive interference, 272–81
Dichroic material, 288
Difference tone, 95, 100
Differential brightness
    discrimination, 255
Differential hue discrimination,
    254
Differential loudness
    discrimination, 96
Differential pitch discrimination,
    96
Diffraction, 266–71
Dispersion, 197, 233–36

Dissonant tones, 78
Doppler effect, 236–39
Double bass, 179
Driving an oscillator, 26
Driving a string, 55
Driving frequency, 26–27, 57–58
Duality, wave-particle, 335

Ear
    Capability, 90–97
    Structure, 97–101
Eardrum, 98
Electromagnetic radiation, *see*
        Waves, Electromagnetic
Emulsion, 194
Energy, 4, 25, 30, 194, 202
Equilibrium position, 19, 21
Eustachian tube, 98
Extraordinary ray, 291
Eye
    Capability, 249
    Comparison to camera, 248
    Defects, 258
    Structure, 255
Eyeglasses, 233, 261

Fan, 33
Fetch, 129
Fixed-frequency instrument, 70, 83
Fluorescence, 203
FM (Frequency Modulation), 324
Focal point, 224
French horn, 187
Frequency, 19, 21, 32, 91, 198
    Ranges of instruments, 184

Friction, 25
Fundamental, 48

Gas discharge tube, 204
Grating, 281
Green flash, 312

Halo, 307
Harmonics, 48, 160
Hertz, 20
Hitting a string, 55, 74, 76
Holography, 324–28
Hue, 250

Ice crystals, 308–10
Ideal continuous string, 48
Image, 194, 229–33
    Real, 229, 232
    Virtual, 230, 233
Incandescence, 202
Incident wave, 148, 215
Infrared radiation, 5, 197, 203
Initial conditions, 50
Intensity, 91, 93
Interface, 217
    Convex and concave, 223
Interference, 271–81
    Thin film, 276
    Two slit, 276

Laser, 318
    Construction, 318–22
    Uses, 322–24
Lenses, 226–29
    Biconcave, 229
    Biconvex, 227
    Ray diagram, 231
Light
    Sources, 202–5
    Speed, 196
    as a Transverse wave, 288
Light pipes, 220
Lissajous figures, 9, 24
Longshore current, 137
Loudness, 91, 93
Loudspeaker, 60

Masking, 96, 100
Mass transport, 137–39
Medium, 4, 234
Melde's experiment, 56
Microphone, 32, 61, 113
Mirage, 311
Monochromatic, 250, 321
Motion
    Deep water waves, 116
    Periodic or cyclic, 19
    Shallow water waves, 117
    Transverse, 45

Musical scale, 70, 77–83
    Fifth, 78
    Fourth, 77
    Just or natural diatonic, 79
    Octave, 70, 77
    Pentatonic, 79
    Pythagorean, 78
    Stretched, 76

Nanometer, 198
Natural frequency, 25–27, 30, 57
Noctilucent clouds, 311
Nodal lines, 59
Nodes, 46, 49, 154, 165
Normal (to an interface), 214
Normal modes, 44–54
    Determining wavelengths and
        frequencies, 162
    Labeling, 46
    as Standing waves, 157–63
    Superposition of, 50–54

Object beam, 325
Oboe, 183
Octave, 70, 77
Optically active material, 292
Ordinary ray, 291
Organ of Corti, 99
Organ pipes, 163
Oscillators
    Bead on elastic string, 45
    Driving, 26
    Resonating, 27
    Responding, 26
    Simple, 3, 7, 21
Oscilloscope, 5, 21, 77
Overtones, 48

Parhelion, 309
Partially reflecting mirror, 320
Pendulum, 3, 12, 22, 25–26
    Bob, 19
    Simple, 18
    Two coupled, 29
Period, 19, 21
Persistance of vision, 6, 260
Phase, phase angle, 22–24
Photons, 335
Piano
    Action, 74
    Bridge, 72
    Keyboard, 70–71
    Pedals, 75
    Sounding board, 72
    String, 71
Pinhole camera, 194
Pitch, 8, 91, 163
Place theory of hearing, 99–100
Plucking a string, 55

Polarization, 284–92
    Diagram, 285
    by Reflection, 289
Polarizer, 286
Position versus time curve, 9
Primary colors
    Additive, 199, 254, and Plate 2
    Artists, 252
    Psychologists, 252
    Subtractive, 200 and Plate 2
Primary rainbow, 302 and Plate 8
Principle of superposition, 34, 236
Prism, 198, 234

Quanta, 335

Rainbow, 300–304 and Plate 8
Rarefaction, 111
Ray, 212
Red sunsets, 304
Reference beam, 325
Reflection, 217–18
    at End of string, 148–51
    Phase change, 150
    Total internal, 218–22
Refraction, 120, 214–17
Resonance, 25, 29, 57, 179–80, 183
Resonant frequency, 27
Response curve, 27, 31, 58, 73, 94,
    253, 256
Rip currents, 140
Rotary motion, 33

Saturation, 250
Saxophone, 183
Scattering, 281–84
    in the Atmosphere, 304–6
Secondary rainbow, 302 and Plate
    8
Seiches, 174–78
Semitone, 79, 81
Shadow, 306
Signal generator, 34, 62
Simple harmonic motion, 49, 61
Sine curve, 7, 8, 21, 23, 31, 34, 49
Slinky, 111
Sound
    as a Longitudinal wave, 112
    Observing on an oscilloscope, 7,
        31–32
    Produced with a loudspeaker, 60
    Speed, 113
Spectrum, 204
Spring-mass system, 19, 24
Standing waves
    Loops, 161, 165
    as Normal modes, 157–63
    Production, 152–57
Stereo system, 13

Stimulated emission, 320
String
    Bowing, 181–82
    Driving, 55
    Ideal continuous, 48
    Linear mass density, 71
    Piano, 71, 75
    Tension, 56, 163
    Violin, 179
Strobe, strobe light, 32–34, 76
Subatomic particles, 335
Sub sun, 310
Sun dog, 309
Sun pillar, 309
Supernumerary bows or arcs, 303
    and Plate 8
Superposition, 34
    to Form standing waves, 152–57
    of Normal modes, 50–54

Tide, 128, 133–34
Timbre, *see* Tone quality
Tone quality, 62, 76, 92, 186–89
Total internal reflection, 218–22
Transmission, 151
Tremulo, 8
Trombone, 187
Trough, 107
Trumpet, 187
Tsunami, 128, 132
Tuning fork, 29–31, 34
Tweeter, 62

Ultraviolet light, 5, 197, 203
Uncertainty, experimental, 20
Undertow, 140

Vibrating systems, *see* oscillators
Vibration parttern, 5, 7–8, 21, 91
    Triangular, sawtooth, and
        square, 34, 62
Vibration recipe, 54, 62, 74, 92,
    186–89
Vibrations
    Connection with waves, 2
    Description of, 18–20
    Electrical, 34, 61
    Measurement of, 20, 31
    Overview, 2
Vibratto, 8
Viola, 179
Violin, 178–82
    Bass bar, 180
    Bow, 181
    Bridge, 179–80
    Sound post, 180
    String, 181

Waveform, *see* Vibration pattern
Wavefront, 117, 212
Wavelength, 109, 196, 198
Waves
    Amplitude, 108
    Converging, 225

Diverging, 223
Electromagnetic, 5, 107, 197
Frequency, 108
Gravity, 114
Height, 115
Light, 5, 107, 194–98
Longitudinal, 111
One-, two-, and
    three-dimensional, 106, 210,
    212
Overview, 2
Period, 108
Plane, 213
Probability, 335
Sound, 4, 107, 196
Speed, 108
    Ideal string, 110
    Sound, 113
    Water, 115–17
Spherical, 213
Steepness, 129
    on a String, 107, 218
Transverse, 111
Traveling, 106
Water, 114–20, 218
Weber's law, 96
Whispering galleries, 222
White light hologram, 327
Whole tone, 79, 81
Wind-generated waves, 128–31
Wind instruments, 183–89
Woofer, 62

David M. Scott is Professor of Physics at The Ohio State University's Mansfield Campus. He received his B.S. in Mechanical Engineering from the Massachusetts Institute of Technology in 1955 and spent eight years working for General Motors and Monsanto. Returning to academic pursuits, he was awarded the Ph.D. from Ohio State in 1969 and then joined the faculty. For over ten years he has taught a course on which this book is based. His research interest is the area of elementary particles.